Fragrance and Flavor Substances

Fragrance and Flavor Substances

Proceedings of the Second International
Haarmann & Reimer Symposium
on Fragrance and Flavor Substances
(New Products, Processes and Aspects of Product Safety)

September 24-25, 1979
New York City

Edited by
Rodney Croteau
Washington State University
Institute of Biological Chemistry
and Biochemistry/Biophysics Program
Pullman, Washington 99164

D & PS. Verlag, 3017 Pattensen 1, W.-Germany
1980

Publishers: D & PS. Verlag, Pattensen
ISBN 3-9800422-0-0

FOREWORD

The First International Symposium on Fragrance and Flavor Substances was held in Bad Pyrmont in 1974 to commemorate Haarmann & Reimer's first one hundred years of service. The dual themes of the First Symposium were "Chemistry and Biochemistry of Fragrance and Flavor Substances" and "Chemical Structure and Biological Effects", themes of very broad scope encompassing many different facets of the fragrance and flavor field. This meeting brought together experts from research areas as diverse as electrophysiology, analytic and synthetic chemistry, and biochemistry, but all having a common interest in fragrance and flavor materials. The First Symposium provided a level of interaction among the attending scientists that was unprecedented in this field. The success of the first meeting encouraged Haarmann & Reimer to convene this conference, the Second International Symposium on Fragrance and Flavor Substances, an event which marks the retirement of Dr. Kurt Bauer as Vice President for Research and Development of Haarmann & Reimer.

The theme of this Second Symposium is "New Products, Processes, and Aspects of Product Safety", a theme of defined focus on topics of crucial importance. New Products and new syntheses are, of course, aspects of long standing interest, for they are the mainstay of the fragrance and flavor industry. The topic of product safety reflects an increasing universal concern for man and his changing environment. Indeed, new products and new applications are now inseparable from health and safety considerations. In this Symposium, the latest advances in the preparative chemistry and the odor characteristics of a wide range of compounds are reviewed. Safety considerations in the broad sense are discussed, and recent work on the mechanisms of cutaneous toxicity is described. Included also is a review of terpene biosynthesis which compliments synthetic studies in this area and underscores the larger point that fundamental biochemical processes are the basis of both odor and taste perception as well as toxicity considerations.

An important role of this Symposium, as was true of the First, is to bring together scientists having a common interest in fragrance and flavor materials in a setting in which the present can be assessed and a perspective of the future formulated. As Chairman, one of my functions is to describe, in brief overview, the unifying concepts of this Symposium, and to provide such a look to the future. At first this might seem like a difficult task for me in that my perceptions are those of a biochemist looking at the biochemical foundations underlying the production, acceptance and safety considerations of a fragrance or flavor product. Thus, on one level, I see the synthesis of a fragrance, such as jasmine, as mimicking a natural biosynthetic process; I regard the perception of an odor, such as a musk, as the interaction of a small molecule with a protein receptor that results in a sensory signal; and I view the toxicity of a potentially useful aromatic chemical as the metabolic inability to deal with a reactive compound. On a more fundamental level, however, I see the economic synthesis of a fragrance or flavor substance, indeed process chemistry in general, in terms of catalysis – the exploitation of the inherent reactivity of molecules. Catalysis is also the essence of biosynthesis – synthesis by enzymes, the protein catalysts. Our understanding of chemical catalysis is far more advanced than is our understanding of biochemical catalysis, and this point is well illustrated by the contents of this Symposium. Yet, significant advances are being made in the enzymology of fragrance and flavor biosynthesis. Continued progress, when coupled with the tremendous strides in molecular biology, can in the foreseeable future provide the basis for industrial scale production of fragrances, flavors and other fine chemicals by enzymatic means. We can also look forward to a greater understanding of the chemistry, biochemistry and physiology of odor and taste perception and to a deeper appreciation of the molecular basis for toxicity, both of which will allow a rational, more integrated approach to the design of effective and safe fragrance and flavor materials. It is symposia of this type that provide a forum for the exchange of ideas among scientists, from which the seeds of such innovations will grow.

Many people, in addition to the speakers and their coauthors, contributed in a significant way to the success of this Symposium. In particular, I wish to thank Peter Weyerstahl for his assistance at the podium, and Ian Gatfield and Lothar F. Kümper and his staff for their outstanding job in maintaining the smooth pace of the sessions. For his encouragement and guidance, I thank Jim Adams, and, for contributions far too numerous to list, I extend my sincerest gratitude to Kurt Bauer. Finally, on behalf of all of the participants, I especially thank Haarmann & Reimer for making this Symposium possible.

Rodney Croteau

CONTRIBUTORS

KURT A. BAUER, Research Department, Haarmann & Reimer GmbH, 3450 Holzminden, West Germany.

CLAUDE BENEZRA, Laboratoire de Dermato-Chimie, Université Louis Pasteur, Strasbourg, France.

RODNEY CROTEAU, Institute of Biological Chemistry and Biochemistry/Biophysics Program, Washington State University, Pullman, WA 99164, U.S.A.

HELMUT FIEGE, Research Department II, Bayer AG, 5090 Leverkusen, West Germany.

DETLEF J. HAGENA, Research Department, Haarmann & Reimer GmbH, 3450 Holzminden, West Germany.

ALFRED K. KÖRBER, Research Department, Haarmann & Reimer GmbH, 3450 Holzminden, West Germany.

ALFRED KREMPEL, Research Department, Haarmann & Reimer GmbH, 3450 Holzminden, West Germany.

HEINRICH KUHLMANN, Department of Organic Chemistry, Technical University Aachen, 5100 Aachen, West Germany.

GERD K. LANGE, Research Department, Haarmann & Reimer GmbH, 3450 Holzminden, West Germany.

ROGER D. MIDDLEKAUFF, Bonner, Thompson, O'Connell and Gaynes, 900 17th St. N.W., Washington, D.C. 20006, U.S.A.

REINER G. MÖLLEKEN, Research Department, Haarmann & Reimer GmbH, 3450 Holzminden, West Germany.

HERMANN J. OEDIGER, Central Research and Development, Bayer AG, 5090 Leverkusen, West Germany.

ANDREAS SCHULZE, Central Research and Development, Bayer AG, 5090 Leverkusen, West Germany.

MERTON V. SMITH, Division of Food and Color Additives, Food and Drug Administration, Washington, D.C. 20204, U.S.A.

HERMANN STETTER, Department of Organic Chemistry, Technical University Aachen, 5100 Aachen, West Germany.

KARLFRIED WEDEMEYER, Research Department II, Bayer AG, 5090 Leverkusen, West Germany.

JOHN A. WENNINGER, Division of Cosmetics Technology, Food and Drug Administration, Washington, D.C. 20204, U.S.A.

PETER K. WEYERSTAHL, Department of Organic Chemistry, Technical University Berlin, D-1000 Berlin 12, West Germany.

CONTRIBUTORS

Prof. Rodney Croteau

Dr. Kurt A. Bauer

CONTRIBUTORS

Prof. Claude Benezra

Dr. Helmut Fiege

Dr. Detlef J. Hagena

Dr. Alfred K. Körber

Dr. Gerd K. Lange

Roger D. Middlekauff

Dr. Reiner G. Mölleken

Dr. Hermann J. Oediger

Dr. Merton Smith

Prof. Hermann Stetter

John A. Wenninger

Prof. Peter K. Weyerstahl

CONTENTS

1. THE BIOSYNTHESIS OF TERPENE COMPOUNDS

R. Croteau*

Summary Monoterpenes and sesquiterpenes are naturally occurring flavor and fragrance substances that are found in the essential oils produced by plants. A brief description of the structure and function of monoterpenes and sesquiterpenes is followed by a review of recent advances in the biochemistry of these compounds, with special emphasis on biosynthetic pathways. The enzyme-catalyzed synthesis of *d*-camphor in sage and of *l*-menthol in mint is described in detail. The origin of sesquiterpenes is discussed with particular reference to the similarities between monoterpene and sesquiterpene biosynthesis.

Introduction: The Essential Oils

It has been known for several thousand years that the flowers, fruits, leaves, barks and roots of many plants contain volatile, odorous substances, collectively called the essential oils. Highly prized by ancient man, the essential oils, in the form of herbs and spices, became one of the earliest recorded items of commerce, and it is unlikely that any other commodity has provided such a stimulus to world exploration and to the development of international trade [1, 2]. From the ancient trade in raw spices, the isolation and preparation of odorous principles from plants, and the synthetic duplication of the essential oils and their constitutents has developed, in modern times, into a large and flourishing industry.

While of great utility to mankind, the classical view has held that essential oils represented inactive "waste products" of plant metabolism and thus possessed no significant biological function [3]. Only recently have we come to realize that the essential oils and their constitutents may be of benefit to the plants which produce them; providing a means of chemical communication by which the plant maintains itself against competitors, predators and pathogens [4, 5]. The ability of essential oils to function as chemical signals appears to stem from their volatility in air (and thus the ability to influence another organism at some distance from the source plant) and from the complexity of composition and structure of the constituents (and thus the ability to transmit a complex and selective biological message). Additionally, recent evidence has demonstrated that essential oil constituents are both rapidly synthesized and catabolized by plants [6], and thus may play a dynamic (though as yet unknown) role in

* Rodney Croteau, Washington State University, Institute of Biological Chemistry and Biochemistry/Biophysics Program, Pullman, Washington 99164 U.S.A.

metabolism. In this context of a functional role, it is important to note that the synthesis and accumulation of essential oils is generally associated with the presence of specialized glandular structures, most often situated on or near the surface of the plant [6] (Fig. 1). Such a location is ideal for release of essential oil constituents to the atmosphere, and is consistent with a chemical defense function. Additionally, the oil glands provide a means of sequestering the essential oils from primary metabolic processes; yet, the sites are not so isolated as to prevent catabolism and the apparent metabolic recycling of the essential oil constituents.

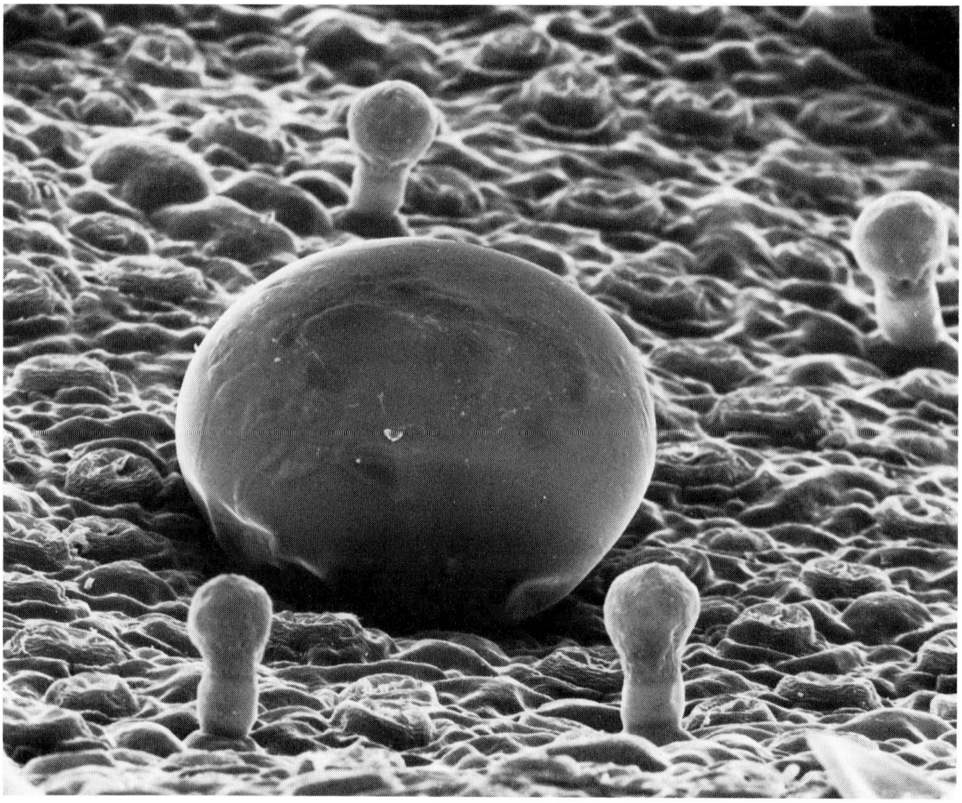

Fig. 1 Scanning electron micrograph showing one large 10-celled oil gland surrounded by several smaller 3-celled oil glands on the upper surface of a spearmint leaf. The diameter of the large gland is about 80 µm. (From Croteau [70]).

Monoterpenes and Sesquiterpenes: General Considerations

Early studies on the chemical composition of the essential oils lead to the discovery of a series of related olefins of formula $C_{10}H_{16}$. These compounds became known as terpenes, derived from the German terpentine (turpentine). Related oxygenated compounds with a ten-carbon skeleton were also isolated, and the crystalline ones became known as *camphors* because of their physical resemblance to true camphor. Thus, there was a "thyme camphor" (thymol) and a "peppermint camphor" (menthol). Later, when further structural similarities were noted, the oxygenated compounds were included under the general heading of terpenes, with the term "camphor" now being restricted to a specific compound. The term "terpene" is now loosely applied to all compounds of terpenoid (isoprenoid) structure (see below).

The family of structurally related ten carbon substances, called monoterpenes, were usually constituents of the lower boiling fractions of essential oils. When attention was later turned to the higher boiling fractions, the presence of similar fifteen carbon containing compounds was revealed. These compounds, the sesquiterpenes, have since been found to be fairly widespread in nature and, because of their volatility, they contribute materially to the odor and flavor of essential oils. Higher terpenes such as diterpenes (C_{20}) and triterpenes (C_{30}) are constituents of extracted plant oils, but these substances are generally eliminated by distillation. It is, therefore, the volatile monoterpenes and sesquiterpenes that constitute the characteristic components of many essential oils, and it is these compounds that are the subject of the present review.

A few examples of common monoterpenes and sesquiterpenes are illustrated in Figure 2. The apparent unit or building block from which the terpenes are constructed is one of five carbon

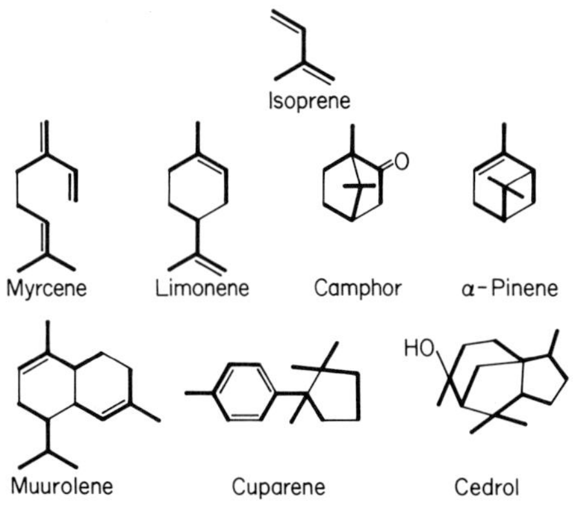

Fig. 2 Representative monoterpenes and sesquiterpenes and their division into "isoprene units".

atoms (i. e., the isoprene unit) and this fact was recognized at a quite early stage of chemical investigation. It was, in fact, the finding that monoterpene hydrocarbons were formally derivable by repetitive joining of the C_5-isoprene unit that led Wallach in 1887 to formulate the "isoprene rule" [7]. This principle–that most terpenoids could be hypothetically constructed by a "head-to-tail" joining of isoprene units–was a major advance in terpenoid chemistry as it provided the first unified concept for a common structural relationship among terpenoid natural products. While isoprene itself was once considered to be the actual biological precursor of terpenes in plants, this notion was discarded because isoprene could not be demonstrated as a natural product (isoprene was shown to be a natural product in 1961 [8]). The smallest terpene and, thus, the "terpene unit" was therefore considered to comprise ten-carbon atoms, from which the designations monoterpene (C_{10}), sesquiterpene (C_{15}), diterpene (C_{20}), etc. originated. Within these broad categories, further divisions are made on the basis of number of carbocyclic rings (acyclic, monocyclic, bicyclic, etc.) and skeletal arrangements (*p*-menthanes, bornanes, fenchanes, etc.).

As the structures of an increasing number of terpenes were determined, it became apparent that not all terpenes conformed to the "isoprene rule" in its simplest form. Thus, some terpenes were not exact multiples of five, but lacked one or more carbon atoms. Some showed a "head-to-head" arrangement of isoprene units rather than the regular "head-to-tail" pattern, while others, because they had undergone rearrangement, did not fit either pattern. To bring such a diverse assortment of terpenoids within a common conceptual framework was a formidable task, which was, however, accomplished by Ruzicka in 1953 when he formulated the "biogenetic isoprene rule" [9]. This hypothesis, which ignores the precise character of the biological precursors and assumes only that they are isoprenoid in structure, is based largely on mechanistic considerations. Ruzicka's general scheme for the origin of monoterpenes is reproduced in Figure 3. In this hypothesis, the neryl (*cis*) cation undergoes cyclization to the monocyclic α-terpinyl cation which, by a series of internal additions, hydride shifts and rearrangements, gives rise to the parent cations of the various structural types. Termination of the ionic reactions is envisioned to proceed by elimination of a proton to give the corresponding olefin, or by addition of a nucleophile (e.g., OH^- from water) to give the corresponding alcohol. Similar schemes, based on cationic mechanisms, were formulated for sesquiterpenes (see below) and for higher terpenes as well.

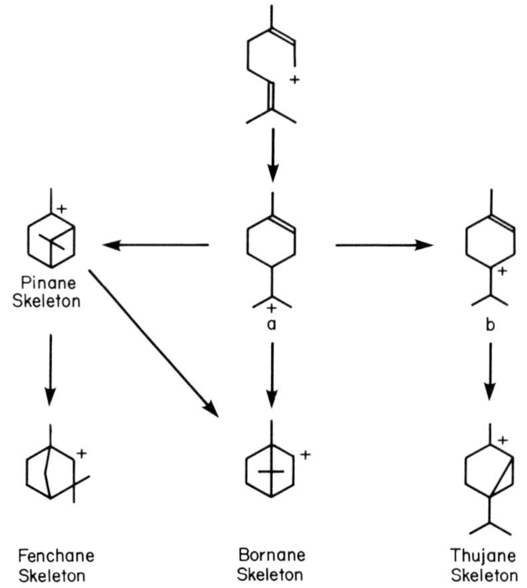

Fig. 3 Ionic scheme for the formation of monoterpenes via the α-terpinyl cation (a) and the terpinen-4-yl cation (b) (after Ruzicka [9]).

While the biogenetic isoprene rule provided a useful working hypothesis for the biosynthesis of mono- and sesquiterpenes, it was not until the discovery of mevalonic acid and its conversion to isopentenyl pyrophosphate (the biological isoprene unit) that an understanding of terpenoid biosynthesis on a molecular level was possible. The enzymatic conversion of mevalonic acid (the first committed precursor of all terpenoids) to isopentenyl pyrophosphate, and the subsequent conversion of this compound to geranyl pyrophosphate and farnesyl pyrophosphate (the first C_{10} and C_{15} compounds, respectively, to arise by this pathway) are illustrated in Figure 4. Detailed discussion of these early steps common to all terpenoid biosynthetic pathways can be found in recent reviews [10–12]. It is important to note here that the first C_{10} isoprenoid, and the precursor of essentially all higher terpenoids, is geranyl pyrophosphate (i.e., the *trans* Δ^2-isomer) (Fig. 4), while on stereochemical grounds it was the *cis* Δ^2-isomer that was proposed as the precursor of cyclic monoterpenes (Fig. 3). Later authors were to propose neryl pyrophosphate as the probable direct precursor of cyclic monoterpenes [13, 14], and to direct considerable effort toward determining how neryl pyrophosphate might be formed in the plant [15]. However, very recent studies with isolated enzyme systems suggest that the direct precursor of cyclic monoterpenes is, in fact, geranyl pyrophosphate. It is this work, as well as other selected examples intended to illustrate the application of modern biochemical techniques to the study of terpene biosynthesis at the enzyme level, that will be described here. For more comprehensive treatment of the subject, the reader is referred to recent reviews on monoterpene [15–19] and sesquiterpene [16, 20–23] biosynthesis.

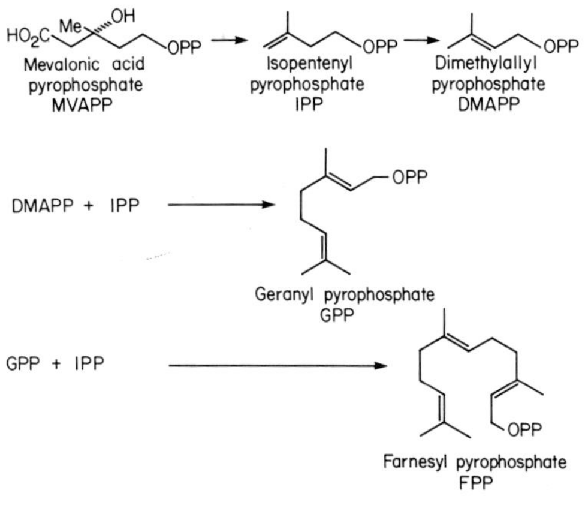

Fig. 4 Enzymatic conversion of mevalonic acid pyrophosphate to geranyl pyrophosphate and *trans,trans*-farnesyl pyrophosphate.

The Origin of Camphor

The main source of natural camphor is the so-called camphor tree *Laurus (Cinnamonum) camphora* of Asia which produces the *d*-isomer. Numerous other species, however, produce camphor, including many of the common herbs and spices. The *d*-stereoisomer appears to be more widely distributed in nature than the *l*-enantiomer. Owing to the ease with which it can be isolated and to its reputed medicinal properties, camphor has been known since prehistory, being first mentioned in early Far Eastern literature [24]. Camphor appears to have been introduced to Europe about the tenth century by Arab traders under the name of *Káfár* or *Kaphur* (from the Sanscrit *Kapūra* meaning "white"). Because of its reported therapeutic properties, camphor was quickly accepted in the West, where much confusion reigned as to the source of this material (it was at one time thought to be a mineral resin). Traders wishing to maintain a monopoly on the supply of camphor no doubt contributed to this confusion. With the establishment of camphor as a natural plant product, the investigation of the chemical and physical properties of this unique material began. The history of camphor is considered to be one of the longest and most involved of any natural product [25].

One of the first mentions of camphor in a chemical context was by the alchemist Libavius [26], who in 1595 described the solution of camphor in nitric acid. The next two-hundred years witnessed numerous studies on the properties of this compound by some of the most prominent chemists of the time, including Boyle and Lavoisier [25]. The composition and molecular weight of camphor were determined by Dumas [27] in 1833, and the formula for camphor as $C_{10}H_{16}O$ established. The next sixty years were required to establish the correct structure of camphor, a difficult problem due to the tendency of camphor and its derivatives to undergo intramolecular rearrangements. During this sixty year period over thirty structures were proposed for camphor (some notable examples are illustrated in Figure 5) culminating in the correct formulation by

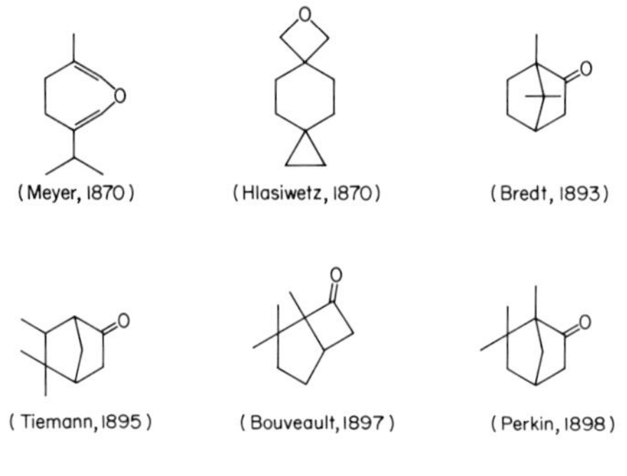

Fig. 5 Early proposals for the structure of camphor.

Bredt [28] which was not, however, generally acknowledged until many years later. With the structure of camphor accepted, many of the rearrangements which had misled earlier investigators were now explained, and several syntheses of camphor were designed. The increasing use of camphor as a plasticizer, particularly in the manufacture of celluloid and explosives, soon lead to a demand that exceeded the supply of natural material, and from the synthetic studies commerical processes for the production of camphor were developed. Figure 6 illustrates the Bakelite-Xylonite (BXL) process [29], which can be seen to utilize α-pinene (the primary constituent of turpentine) as the starting material and to proceed through a reaction

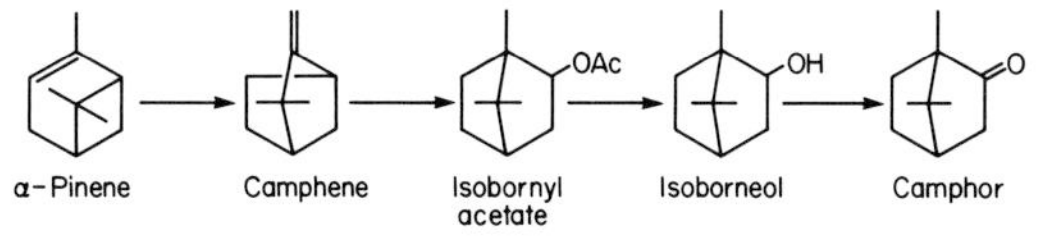

Fig. 6 Bakelite-Xylonite process for the synthesis of camphor (from Bean [29]).

sequence involving no fewer than three Wagner-Meerwein rearrangements. It was, in part, the rearrangements involved in the synthesis of camphor that formed the basis of studies by Wagner [30] and Meerwein [31] from which this fundamental rearrangement process derived its name. The racemic end-product of the BXL process can be purified to 98% by sublimation or crystallization, and the world usage of this product is probably somewhat in excess of 400 tons yearly [29]. While the industrial significance of camphor has declined over the years, the chemistry of this unique material continues to provide challenge to the modern day researcher [32–34].

With Wallach's formulation of the isoprene rule [7] and the formulation of the correct structure of camphor [28], it was immediately realized that camphor was "isoprenoid" or "terpenoid", and thus biogenetically related to other members of this rapidly growing family of natural products. It was Ruzicka, however, who in formulating the biogenetic isoprene rule [9] provided a theoretical framework for the biosynthetic construction of the bornane (camphane) monoterpenes via the cyclization of the neryl cation (see Fig. 3). Later workers were to suggest neryl pyrophosphate as the likely biological precursor [13, 14], and in the early 1970's two groups [35, 36] showed that the labelling pattern of camphor derived from specifically labeled precursors in intact plants was consistent with Ruzicka's cyclization scheme. The first enzymatic synthesis of camphor in a cell-free preparation from leaves of sage (*Salvia officinalis)* was demonstrated in 1976 [37]. This soluble enzyme system converted the acyclic precursor [1-^{3}H]neryl pyrophosphate to *d*-[3-^{3}H]borneol in the presence of Mg^{++}, and in the presence of NAD, dehydrogenated the borneol formed to *d*-[3-^{3}H]camphor (Fig. 7). The location of the tritium in borneol and camphor indicated a direct conversion of substrate to products, and resolution of the biosynthetic products showed them both to be the *d*-isomer, the same stereoisomers produced *in vivo* [38]. A specific search of the enzymatic reaction products for the diastereomeric *l*-isoborneol revealed that this alcohol was not formed in detectable yield. Small amounts of the monocyclic alcohol α-terpineol were formed by the crude cell-free preparation;

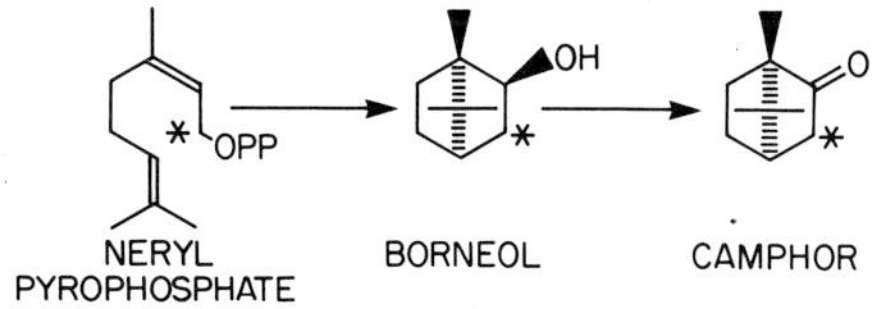

Fig. 7 Pathway for the enzymatic conversion of [1-^{3}H]neryl pyrophosphate to *d*-[3-^{3}H]borneol and *d*-[3-^{3}H]camphor. The asterisk indicates the position of tritium in the structures, which are stereochemically correct.

however, subsequent studies showed that α-terpineol was not an intermediate in borneol biosynthesis [37]. This was the first report on the enzymes involved in the biosynthesis of camphor, and the results suggested that camphor was derived by the cyclization of neryl pyrophosphate to borneol followed by dehydrogenation of this bicyclic alcohol (Fig. 7).

Attempts to purify the *d*-borneol synthetase activity by gel permeation chromatography resulted in the apparent loss of catalytic capability; however, subsequent recombination of column fractions demonstrated that two clearly separable enzymatic activities (eluting in the molecular weight regions corresponding to 95,000 and 50,000) were required for the conversion of neryl pyrophosphate to borneol [38]. Several lines of evidence indicated that a water-soluble, dialyzable intermediate (produced by the higher molecular weight component) was involved in this transformation. The intermediate was isolated and conclusively identified as *d*-bornyl pyrophosphate [38]. Thus, *d*-borneol appeared to be formed by a pathway involving two discrete enzymatic steps: the cyclization of neryl pyrophosphate to *d*-bornyl pyrophosphate (by the 95,000 MW protein component) and the hydrolysis of *d*-bornyl pyrophosphate to *d*-borneol (by the 50,000 MW protein component) (Fig. 8). The intermediate role of *l*-bornyl pyrophosphate in the biosynthesis of *l*-borneol has also been demonstrated in soluble enzyme preparations from rosemary (*Rosmarinus officinalis*), indicating that this pathway for borneol formation is probably widespread in the plant kingdom [38]. While cyclic pyrophosphates such as copalyl pyrophosphate, presqualene pyrophosphate, and prephytoene pyrophosphate are involved in the biosynthesis of diterpenes, triterpenes, and tetraterpenes, respectively, the discovery of bornyl pyrophosphate was the first report of a cyclic pyrophosphorylated intermediate in the biosynthesis of monoterpenes. Furthermore, unlike the other cyclic pyrophosphates cited, bornyl pyrophosphate is unique in that the pyrophosphate moiety of the acyclic precursor migrates during the cyclization reaction.

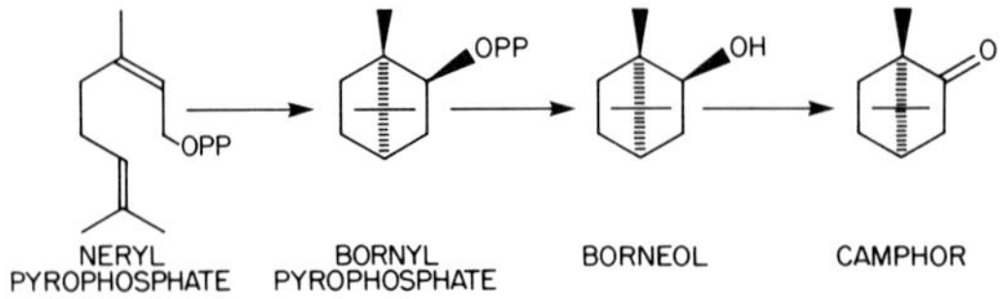

Fig. 8 Pathway for the biosynthesis of *d*-borneol and *d*-camphor via *d*-bornyl pyrophosphate.

Before discussing bornyl pyrophosphate synthetase, which is the key enzyme of the process, it is necessary to review studies with the other requisite enzymes of *d*-camphor biosynthesis; the bornyl pyrophosphate hydrolase and the borneol dehydrogenase. Detailed examination of the sage leaf enzyme preparation revealed the presence of two bornyl pyrophosphate hydrolase activities, one that eluted in the 50,000 MW range (as mentioned above) and one (which was much less active) that eluted in the 20,000 MW range. The higher MW activity was further separated into a pyrophosphate hydrolase (pyrophosphate to monophosphate) and a monophosphate hydrolase (monophosphate to free alcohol) by chromatography on hydroxylapatite. The other hydrolase (20,000 MW) contained pyrophosphate and monophosphate hydrolase activities which were inseparable by this or any other chromatographic technique tested. The lower molecular weight phosphohydrolase with both activities was shown to be rather specific for bornyl pyrophosphate [39], and this observation, along with other data [40], suggested that this enzyme was responsible for the hydrolysis of bornyl pyrophosphate *in vivo*. Examination of the other phosphohydrolases indicated that, while bornyl derivatives were hydrolyzed, phosphate and pyrophosphate esters of primary allylic alcohols, such as geraniol and nerol, were the preferred substrates [39]. Of particular significance was the pyrophosphate hydrolase, which was shown to hydrolyze geranyl pyrophosphate to geranyl monophosphate at faster rates than it hydrolyzed neryl pyrophosphate to neryl phosphate. This enzyme, which destroys the substrates for monoterpene biosynthesis, was later proved to be a serious interference in studies on the bornyl pyrophosphate synthetase (see below).

The final step in the biosynthesis of *d*-camphor in sage is the oxidation of *d*-borneol, and the soluble dehydrogenase that catalyzes this conversion was isolated from sage leaf homogenates and resolved from alcohol (ethanol) dehydrogenase [41]. The partially purified enzyme had a molecular weight of about 91,000, exhibited a pH optimum at about 8.0, and preferred NAD over NADP as cofactor. The kinetic properties of the enzyme, as well as the stereochemistry of hydride transfer (B-type), were also examined [41]. In addition to *d*-borneol oxidation, the enzyme readily catalyzed the oxidation of *l*-3-neoisothujanol to *l*-3-isothujone. Several types of evidence indicated that this same enzyme catalyzed the key oxidation step in the biosynthesis of both *d*-camphor and *l*-3-isothujone, which are the major monoterpenes of sage. This last observation prompted us to suggest that monoterpenol oxidations may, in general, be catalyzed by dehydrogenases which are specific for the particular types of monoterpenes produced by the plant [41], a suggestion supported by subsequent studies with other species [42].

Preliminary studies with bornyl pyrophosphate synthetase were carried out using neryl pyrophosphate as the substrate because this precursor was usually more efficient than the *trans*-isomer, geranyl pyrophosphate. However, the apparent substrate preference of bornyl pyrophosphate synthetase varied considerably from preparation to preparation. This very puzzling observation was clarified with the discovery of the pyrophosphate hydrolase (described above) that preferentially hydrolyzed geranyl pyrophosphate to the monophosphate, thus preferentially destroying this substrate. It was the variation in the level of this competing enzyme in the crude preparations that produced the variation in substrate preference. To examine the bornyl pyrophosphate synthetase in any detail, it was therefore necessary to remove the competing phosphatase. This was accomplished by combination of gel filtration on Sephadex G-150, chromatography on hydroxylapatite, and chromatography on *O*-(diethylaminoethyl)-cellulose [43]. In these fractionation steps, activities for the cyclization of neryl pyrophosphate and geranyl pyrophosphate to bornyl pyrophosphate were coincident, implying that the same enzyme used both substrates. On the progressive removal of the competing phosphatase, the synthetase showed an increasing preference for geranyl pyrophosphate as substrate. After the complete removal of the phosphatase, the K_m for geranyl pyrophosphate was 1.4 μM with V_{max} of 11.4 nmol/hr-mg, while with neryl pyrophosphate the K_m and V_{max} were 18 μM and 6.9 nmol/hr-mg, respectively [43]. The preference for geranyl pyrophosphate was very clear under conditions where no interconversion of geranyl and neryl pyrophosphates was detected. The partially purified bornyl pyrophosphate synthetase had an apparent molecular weight of 95,000 and required Mg^{++} for catalytic activity. The enzyme exhibited a pH optimum at 6.2, and was strongly inhibited by the thiol-directed reagent *p*-hydroxymercuribenzoate and by the active serine directed reagent diisopropylfluorophosphate.

Bornyl pyrophosphate synthetase was the first monoterpene synthetase to be isolated free from competing phosphatases, and the first to show a strong preference for geranyl pyrophosphate as substrate. Other monoterpene cyclase preparations show an apparent preference for neryl pyrophosphate as substrate [44–46], while at least an equal number have been shown to utilize both geranyl pyrophosphate and neryl pyrophosphate with roughly equal efficiency in the absence of interconversion of these substrates [46, 47, 48]. However, in none of these cases was the removal of all competing phosphohydrolases assured, and in several instances only crude preparations were utilized. Thus, in these cases, it is uncertain whether the specificities observed are those of the monoterpene synthetases, or whether they reflect the presence of competing phosphohydrolases. The findings with bornyl pyrophosphate synthetase, and recent studies with fenchol synthetase from fennel which also prefers geranyl pyrophosphate [49], indicate that the traditionally proposed role of neryl pyrophosphate as an obligate precursor of cyclohexanoid monoterpenes (for stereochemical reasons) may need to be reconsidered. Indeed, the present controversy surrounding the possible origins of neryl pyrophosphate as a required precursor of cyclic monoterpenes [15] may be of little significance.

The utilization of geranyl pyrophosphate as the natural immediate precursor necessitates reconsideration of the mechanistic role of the cyclase enzymes in the biosynthesis of cyclic monoterpenes. As any mechanism for the formation of bornyl pyrophosphate must now accommodate geranyl pyrophosphate as the preferred substrate, a reaction sequence in which the *trans*-isomer must be transformed to an intermediate stereochemically suitable for cyclization is implied. The reaction sequence must also account for the appropriate stereochemistry of the cyclized product and for the fact that the bornane skeleton arises as the pyrophosphate derivative. With these considerations in mind, a mechanism for the biosynthesis of bornyl pyrophosphate was proposed [43] (Fig. 9). In order to convert geranyl pyrophosphate

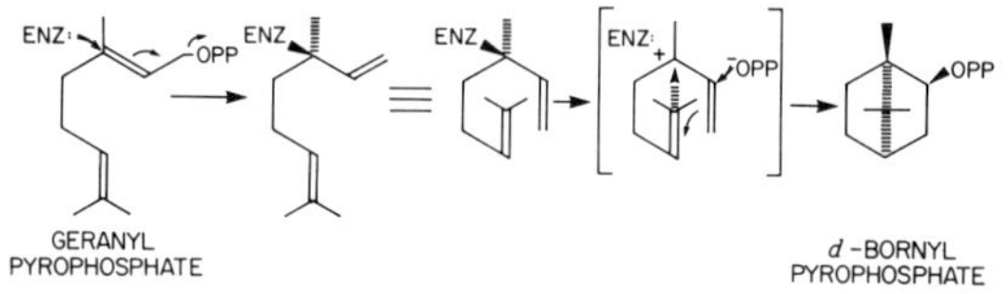

Fig. 9 A possible mechanism for the cyclization of geranyl pyrophosphate to *d*-bornyl pyrophosphate.

into a form stereochemically capable of cyclization, it must first react with the enzyme, with concomitant loss of pyrophosphate, to afford a bound linaloyl intermediate (reaction of neryl pyrophosphate with the enzyme in a similar manner accounts for the activity of this *cis*-isomer). An enzyme bound linaloyl structure would permit rotation around the C2–C3 bond which enables the geranyl pyrophosphate precursor to attain the conformation appropriate for cyclization to bornyl pyrophosphate. While a syncronous cyclization is implied here, this mechanism can be seen to resemble Ruzicka's multistep cationic scheme (Fig. 3) for the generation of the bornane nucleus. The role of the enzyme is, thus, to convert geranyl pyrophosphate to a form appropriate for cyclization, to direct the orientation of entering and leaving groups and thereby achieve the correct stereochemistry, and to provide for the return of the pyrophosphate moiety as the terminating nucleophile. Further details of this hypothetical enzymatic mechanism, including the proposed function of the magnesium cation and the role of a serine residue in binding the pyrophosphate, are described elsewhere [43].

While the mechanism of Figure 9 is consistent with the available data, other feasible mechanisms can be written. Thus, for example, the enzyme bound linaloyl structure could undergo cationic cyclization to an α-terpinyl structure (Fig. 10). Subsequent attack of the tertiary ionic center of the α-terpinyl ion on the trisubstituted double bond results in a stabilized pinyl cation. Binding of this pinyl cation could specifically promote Wagner-Meerwein shift to

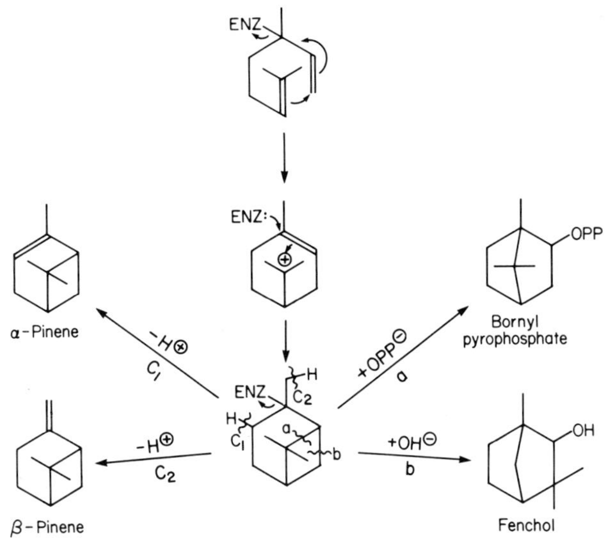

Fig. 10 A possible mechanism for the synthesis of bornane, fenchane and pinane monoterpenes from the same type of intermediate. (Stereochemical considerations are ignored.)

afford a bornyl cation, with the reaction being terminated by pyrophosphate to yield bornyl pyrophosphate. Alternatively, Wagner-Meerwein shift of the methylene group would yield a fenchyl cation, with the reaction being terminated by a hydroxyl ion (from water) to yield *endo*-fenchol. (This mechanism has been proposed to account for the cyclization of geranyl pyrophosphate to *l-endo*-fenchol by an enzyme from fennel [49].) Finally, loss of either of two protons, with double bond formation, would yield the pinenes. This latter type of termination reaction is probably common to the formation of all monoterpene olefins. This overall mechanism, which assigns a central role to the bound pinyl cation, is highly speculative, but perhaps is a more general model in that it accounts for three different families of cyclic monoterpenes and illustrates three different types of termination steps.

While it seems likely that the conversion of geranyl pyrophosphate to a form stereochemically suitable for cyclization is a common first step in monoterpene cyclizations, the remaining features of the cyclization reactions are open to question. To distinguish between the various possibilities will require far more detailed studies of the monoterpene synthetase enzymes. Although no stereochemistry is indicated in Figure 10, this mechanism can be formulated to account for the generation of either naturally occurring enantiomer of the bornane, fenchane, or pinane series. Whether this or another type of mechanism should eventually prove correct, defining how the enzyme actually controls the stereochemistry will certainly prove to be one of the most interesting aspects of monoterpene cyclization reactions.

The Origin of *l*-Menthol and its Congeners

Menthol was known in Asia over 2000 years ago, but the first mention of this compound in the Western world is attributed to Gaubius, who in 1771 separated this substance from peppermint oil [50]. As was the case with *d*-camphor, the chemistry of *l*-menthol is of great historical interest, particularly as regards the development of stereochemistry. *l*-menthol (and its stereoisomers) occurs in nature primarily in the essential oils of *Mentha* species. It is the characteristic component of the oil of peppermint (*Mentha piperita* L.) and is responsible for the cooling sensation associated with this product. Menthol and its stereoisomers are mainly found in the free state, but a number of recent reports have described the natural occurrence of glycosides of these monoterpene alcohols [51–53]. In 1833, Dumas established that menthol had the empirical formula $C_{10}H_{20}O$, but it was another 50 years before it was fully established that menthol was a saturated secondary alcohol [54, 55]. Elucidation of the structure of menthol is generally attributed to Beckmann [56] and Semmler [57], and is based on the observation that pulegone, menthone and thymol all gave menthol on reduction (Fig. 11). That the modern

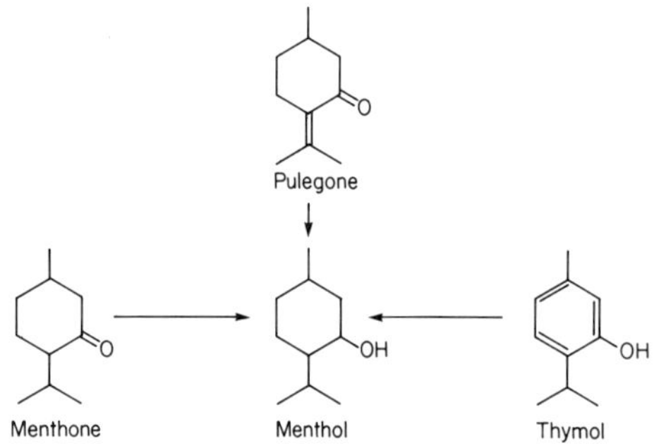

Fig. 11 Some chemical reductions yielding menthol.

Haarmann and Reimer process for *l*-menthol production is based on the hydrogenation of thymol to menthols [58] attests to the importance of this reaction. Inspection of menthol indicates the presence of three asymmetric carbons and thus the existence of eight optically active isomers and four corresponding racemates designated as menthols, neomenthols, isomenthols, and neoisomenthols. Menthone, containing two asymmetric centers, therefore exists as four stereoisomers and two racemates designated menthone and isomenthone. The preparation of the pure stereoisomeric menthols and menthones and the elucidation of their configurations was a very difficult problem shared by many workers including Beckmann, Wallach, Pickard and Read. Accounts of this work, which has contributed in large part to the theoretical development of modern concepts of stereochemistry, are provided by Simonsen [59] and Orloff [60].

With knowledge of the previously mentioned chemical interconversions of pulegone, menthone, and menthol, and the observation that these and other structurally related monoterpenes were present in peppermint oil, Kremers [61] and Reitsema [62] proposed a biosynthetic sequence for the formation of menthol (Fig. 12). Subsequent investigations with peppermint, primarily by Loomis and associates [63–66], have confirmed this scheme, and the evidence for these reactions comes from several types of experiments. Analysis of the monoterpenes from different leaves of the plant showed a predominance of piperitenone and pulegone in the youngest leaves, with a trend toward menthone and then menthol, in the oldest leaves. When $^{14}CO_2$ was administered to the plants, the unsaturated ketones were labeled first, followed by menthone, and later by menthol. Several of the reactions were demonstrated directly by feeding labeled monoterpenes to peppermint leaves. Loomis was also the first to take into account stereochemical considerations in the biosynthetic conversion of piperitenone to *l*-menthol, and this group has recently demonstrated, in cell-free preparations, the stereospecific reductases involved in this sequence of reactions [67]. By contrast to the biosynthesis of camphor, where asymmetry is introduced in the cyclization step, in the formation of *l*-menthol

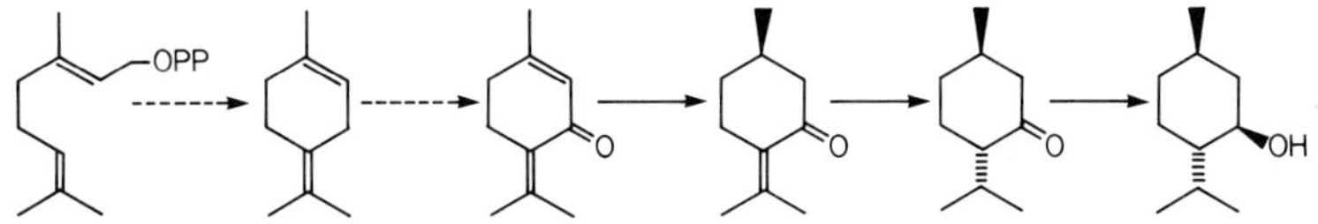

Fig. 12 Pathway for the formation of *l*-menthol in peppermint. Broken arrows indicate postulated reactions.

molecular asymmetry is progressively introduced in three-steps from the symmetric precursor piperitenone. The origin of piperitenone is still uncertain, and this proposed series of reactions is illustrated by broken arrows in Figure 12. The direct conversion of acyclic precursors to terpinolene has been demonstrated [68], but virtually nothing is known about the cyclase enzyme involved in this reaction. Similarly, the mechanism of the oxygenation of terpinolene to piperitenone has not yet been determined. While hydroxylation to the corresponding alcohol followed by dehydrogenation to the ketone might be involved, the ready autooxidation of terpinolene to piperitenone [69] also suggests the possible involvement of hydroperoxides in the reaction.

It has long been known to commercial producers of peppermint that the composition of the oil is greatly influenced by environmental factors. Thus, commercially acceptable peppermint oil can be grown only in a few areas, and in these regions the composition of the oil varies from one season to the next. The effects of environmental factors (such as temperature, photoperiod and light intensity) and agricultural practices (such as irrigation method and fertilizer regime) on the compostion and yield of peppermint oil have been examined [64, 70, 71], and changes in oil character correlated with the physiological condition of the plants. Superimposed on environmental variation, and variation imposed by agricultural practices, is the metabolic process termed "oil maturation", whereby menthone is converted to menthol to produce an oil of acceptable quality. This process, which takes place in peppermint grown under a variety of conditions, is especially prominent in mint grown in the Yakima valley of Washington where maturation occurs at about the time the plants flower (Fig. 13). Thus, unlike *de novo* biosynthesis of monoterpenes, maturation takes place relatively late in the growing season and is coincident with an overall loss of oil content (Fig. 13).

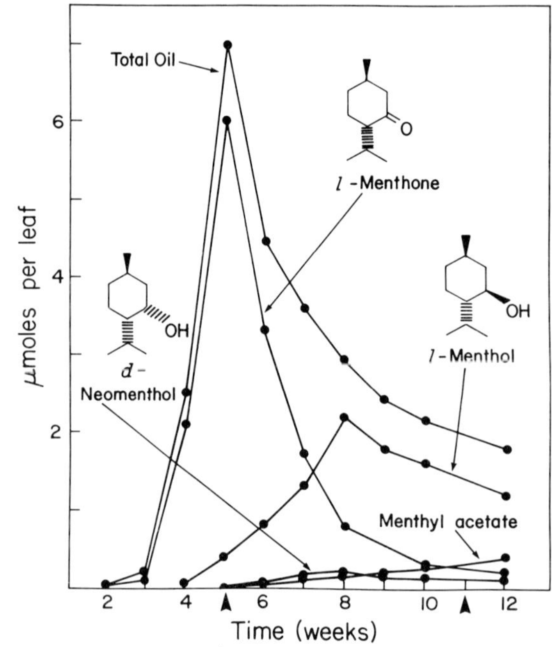

Fig. 13 Essential oil composition of midstem peppermint leaves as a function of growth. The first arrow indicates the approximate time of floral initiation, and the second arrow indicates the approximate time of full bloom. Menthofuran and pulegone are not plotted, although low levels of these compounds were detected from 3 to 6 weeks (from Croteau and Martinkus [51]).

A notable feature of maturation is the large decrease in menthone content that accompanies this metabolic process. This loss in menthone cannot be accounted for by the increase in menthol or other oil constitutents, nor can this loss be rationalized as due to evaporation. Thus, a catabolic process is implied. When *l*-[G-^{3}H]menthone was administered to leaf discs from flowering peppermint plants and the oil then analyzed, it was found that menthone was converted to *l*-menthol, to a small amount of *l*-menthyl acetate, and to a trace of *d*-neomenthol (all expected products from menthone) [51]. However, the bulk of the tritium label from

[G-^{3}H]menthone was not present in the essential oil but rather was located in the water-soluble, non-volatile fraction, thus giving the first indication that menthone was converted to a metabolite other than an essential oil component [51].

Analysis of the water-soluble fraction indicated the presence of two labeled components derived from [G-^{3}H]menthone. The minor component was tentatively identified as a highly oxygenated *p*-menthane monoterpene, while the major metabolite was conclusively identified as *d*-neomenthyl-β-D-glucoside [51]. Menthyl glucoside was recently identified in Japanese mint (*Mentha arvensis*) [52, 53], but the amount of menthyl glucoside formed in peppermint (*Mentha piperita*) was negligible compared to the neomenthyl glucoside. This was the first report on the occurrence of a neomenthyl glycoside in nature, and the first evidence implicating glycosylation as an early step in monoterpene catabolism [51]. Further chemical investigations and studies with $^{14}CO_2$ as a precursor clearly indicated that, during the maturation process, *l*-menthone is reduced to roughly equal amounts of *l*-menthol and *d*-neomenthol [51]. The bulk of the menthol formed appears in the volatile oil (as such, or as the acetate ester) whereas the bulk of the neomenthol formed is converted to the glucoside. The pathways are highly specific in that only menthyl acetate (with little neomethyl acetate) and only neomenthyl glucoside (with little menthyl glucoside) are formed (Fig. 14). Two possible explanations for this specificity seemed

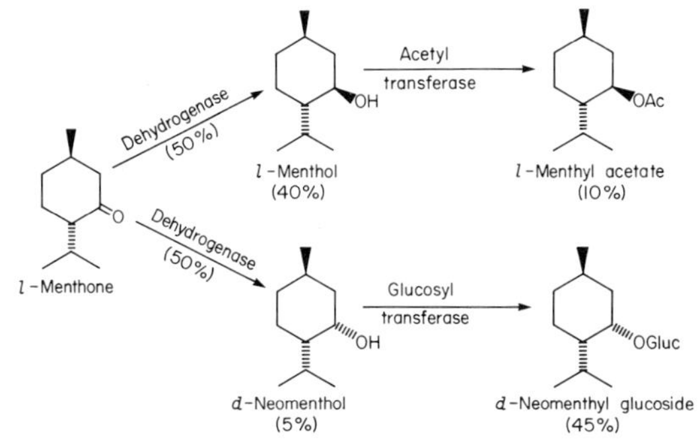

Fig. 14 Pathway for the metabolism of *l*-menthone during maturation in peppermint. The approximate composition of the products derived from menthone is given.

evident: either the transacetylase and the transglucosylase are highly specific for menthol and neomenthol, respectively, or a dehydrogenase specific for menthol production is compartmentized with a transacetylase, while a dehydrogenase specific for neomenthol formation is accessible only to a glucosylating system. When *l*-[G-^{3}H]menthol and *d*-[G-^{3}H]neomenthol were separately administered to peppermint leaf discs, equal amounts of *l*-menthyl acetate and *d*-neomenthyl acetate were formed, and equal amounts of *l*-menthyl glucoside and *d*-neomenthyl glucoside were formed [72]. Thus, specificity of product formation (as in Fig. 14) was observed when *l*-[G-^{3}H]menthone (or $^{14}CO_2$) was the precursor, but not when the alcohols were administered directly. These results strongly suggested that specificity is determined at the menthone reduction step, that the system is compartmentized, and that the transacetylase and the transglucosylase are not highly selective enzymes. This latter suggestion was confirmed by isolating the enzymes involved and examining their properties. Thus, the acetyl CoA-dependent transacetylase utilized *l*-menthol and *d*-neomenthol with nearly equal efficiency in the synthesis of the corresponding acetate [73]. Similarly, the UDP-glucose-dependent transglucosylase exhibited nearly identical K_m and V_{max} for *l*-menthol and *d*-neomenthol, synthesizing both β-D-glucosides with equal proficiency [72]. All of the above results pointed to the enzymatic reduction of *l*-menthone as the key step in the maturation process, and homogenates of leaves from flowering peppermint plants were shown to catalyze the enzymatic reduction of *l*-menthone to nearly equal quantities of *l*-menthol and *d*-neomenthol, with NADPH as the required cofactor (Martinkus and Croteau, unpublished). The activity was localized in the soluble enzyme fraction and, by combination of gel permeation chromatography and affinity chromatography, two distinct and stereospecific dehydrogenases were isolated (one for reduction of menthone to menthol, the other for reduction to neomenthol). The two enzymes have a number of similar characteristics, but differ in kinetic properties suggesting that their activities may be differentially regulated *in vivo.* While the two dehydrogenases are "operationally" soluble enzymes, the tissue extraction procedure employed was quite harsh and the integrity of cellular compartments could easily have been destroyed, thereby causing release of their contents to the soluble phase. Thus, while the evidence cited above clearly implicates compartmentation of pathways as an essential feature of the maturation process, the cellular location of the stereospecific dehydrogenases and their auxiliary enzymes (the tranacetylase and transglucosylase) is a key question that remains to be answered.

The role and fate of neomenthyl glucoside are not presently known. However, if during this apparent catabolic phase the monoterpenes are utilized at sites other than the epidermal oil glands, then a means of transporting these lipophilic materials would be required. Monoterpenyl glycosides, of which many have recently been identified [51], may represent such a transport form. When *l*-[G-^{3}H]menthone is administered to leaves of intact flowering peppermint plants, the neomenthyl glucoside formed in the leaves appears to be specifically transported to the roots, where at least a portion of this material is further metabolized to unidentified products [51, 72]. This observation supports the suggestion that monoterpenyl glycosides are transport derivatives, but much more work will need to be done before gaining a clear understanding of the physiological significance of this observation. Further studies on the properties, regulation, and localization of the enzymes from peppermint, which may be a model for other plants as well, should lead to a better understanding of, and perhaps a rationale for, monoterpene catabolism.

Comparative Aspects of Monoterpene and Sesquiterpene Biosynthesis

As the monoterpenes and sesquiterpenes are commonly found together in essential oils, there is a natural tendency to regard these two types of compounds as closely related. Similarities and differences in occurrence, function and sites of synthesis have been described [6, 16], but the comparative biochemistry of these two families of terpenes is still largely a matter of speculation. The sesquiterpenes contain one more isoprene unit than monoterpenes, allowing much greater flexibility in the construction of sesquiterpenoid compounds. Almost two hundred different carbon skeletons containing up to four carbocyclic rings are found in this terpenoid group, and several thousand individual sesquiterpenes have been identified [74, 75] (this compares to less than five hundred monoterpenes [75]). The sesquiterpenes are the largest single class of terpenoids, representing a tremendous diversity of structure. The bulk of the sesquiterpenes (like the examples in Fig. 2) are constructed by the typical head-to-tail fusion of isoprene units, or can be regarded as rearrangement products of more typical head-to-tail structures.

The sesquiterpenes were considered by Ruzicka [9] to originate from the corresponding cation of *trans*-Δ^2, *trans*-Δ^6-farnesol or, where steric considerations appeared to apply, from the *cis*-Δ^2, *trans*-Δ^6-isomer; the *trans, trans*-isomer yielding the ten and eleven membered ring intermediates, and the *cis, trans*-isomer yielding, additionally, six and seven membered ring intermediates (Fig. 15). Within this theoretical framework, transformations of the monocyclic intermediates could involve further cyclizations by addition of cationic centers to the remaining double bonds, hydride shifts and rearrangements, governed by either electronic or steric considerations [22, 76, 77]. Thus, while the sesquiterpenes are structurally more complex than the monoterpenes, the mechanistic rationale for the formation of both classes of compounds is similar. While Ruzicka's original biogenetic isoprene rule was formulated in terms of the farnesols as precursors, the corresponding *trans, trans*- and *cis, trans*-farnesyl pyrophosphates are now considered to be the more likely direct progenitors of the sesquiterpenes. Most recently, nerolidyl pyrophosphate has been proposed as an intermediate in the formation of certain sesquiterpenes [22, 23, 78]. The proliferation of suggested precursors for sesquiterpenes stems in

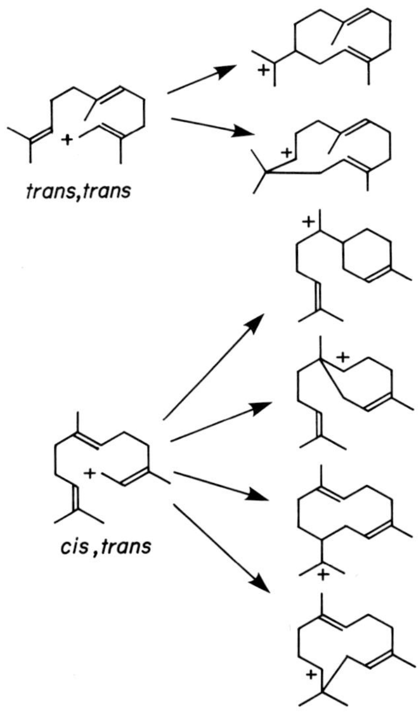

Fig. 15 Hypothetical cyclization of the *trans,trans*- and *cis,trans*-farnesyl cations to monocyclic sesquiterpene intermediates.

large part from the fact that, while *trans, trans*-farnesyl pyrophosphate is the first C_{15} compound to arise in the normal terpenoid pathway (Fig. 4) and is the precursor of most higher terpenes, it is not sterically possible for this *trans, trans*-compound to cyclize directly to all common sesquiterpene structural types. Thus, for example, *cis, trans*-farnesyl pyrophosphate and nerolidyl pyrophosphate have been postulated as precursors of structural types with six- and seven-membered rings (as in Fig. 15). This situation is analogous to that in monoterpene biosynthesis where neryl pyrophosphate has traditionally been regarded as the most likely precursor of cyclohexanoid monoterpenes. The recent finding that geranyl pyrophosphate is, in fact, the preferred direct precursor of several cyclohexanoid monoterpenes (see earlier discussion) therefore suggests that the ubiquitous *trans, trans*-farnesyl pyrophosphate could, by analogy, serve as the direct precursor of sesquiterpene types previously postulated to arise from more "sterically suitable" precursors. Thus, by direct analogy to the proposed cyclization of geranyl pyrophosphate (Figs. 9 and 10), preliminary enzymatic attack on the C_{15} precursor could involve the intermediate formation of an enzyme-bound nerolidyl cation (Fig. 16), which could cyclize through conformer *a* (equivalent to *trans, trans*-farnesyl pyrophosphate) or conformer *b* (equivalent to *cis, trans*-farnesyl pyrophosphate). The configuration at the enzyme bound carbon, the conformer cyclized, the orientation of the isopropylidene group and the conformation of the remaining chain could explain observed stereochemical preferences, and no other precursor than *trans, trans*-farnesyl pyrophosphate need be invoked. While studies at the enzyme level will be required to examine the above hypothesis, some preliminary *in vivo* studies support the concept [79], and others [22] can be rationalized on this basis.

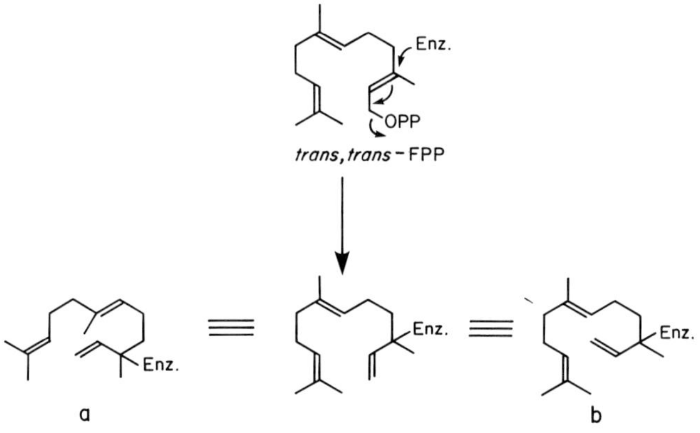

Fig. 16 Possible enzyme-bound intermediates in sesquiterpene biosynthesis.

While a large number of sesquiterpenes have no direct monoterpene counterparts, there are many that are, structurally, the C_5-prenylated analogs of common monoterpenes. In fact, the direct sesquiterpene analogs of virtually all common acyclic, monocyclic, and bicyclic monoterpenes are known, and several examples are illustrated in Figure 17. Similarly, many sesquiterpenes can be regarded as macrocyclic analogs of common monocyclic monoterpenes.

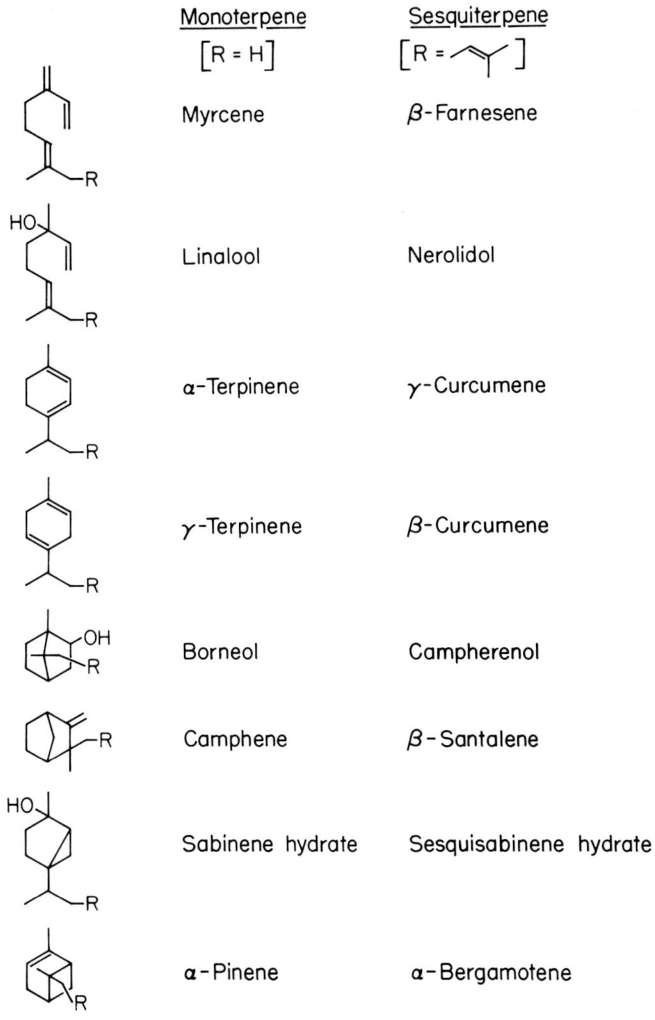

Fig. 17 Some monoterpenes and their sesquiterpene analogs.

One of the simplest families of macrocyclic sesquiterpenes is that based on the germacrane skeleton, and the hypothetical origin of germacrene A, germacrene B and hedycaryol from *trans, trans*-farnesyl pyrophosphate (Fig. 18) can be seen to resemble similar schemes for the formation of the monoterpenes limonene, terpinolene and α-terpineol. Hydride shift and proton loss to yield germacrene C and germacrene D resemble similar processes thought to be involved in the biogenesis of the monoterpene dienes α-terpinene, γ-terpinene and β-phellandrene. Thus, while there is little evidence to support the suggestion, similar

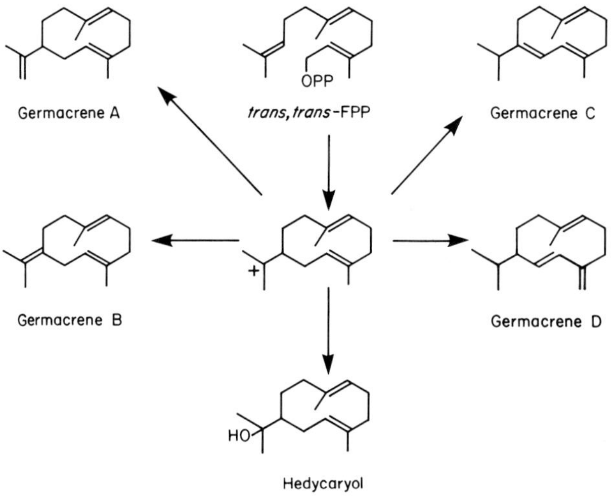

Fig. 18 Postulated origin of germacrenes and hedycaryol from *trans,trans*-farnesyl pyrophosphate.

biosynthetic mechanisms (and probably similar enzymes) for the formation of these related monoterpenes and sesquiterpenes are likely. It is even tempting to speculate that the same

Table 1. Comparison of monoterpene and sesquiterpene biosynthesis by cell-free enzyme preparations

Source of enzyme preparation	Monoterpene (% incorp. of GPP)	Sesquiterpene (% incorp. of FPP)
Salvia officinalis	α-pinene (2.3%)	α-bergamotene (< 0.05%)
Salvia officinalis	bornyl pyrophosphate (as borneol) (4.5%)	campherenol (< 0.05%)
Thymus vulgaris	γ-terpinene (3.7%)	β-curcumene (0.13%)
Chenopodium ambrosioides	α-terpinene (5.8%)	γ-curcumene (0.19%)
Carum carvi	β-phellandrene (0.11%)	germacrene D (1.2%)

GPP = [1-^{3}H]geranyl pyrophosphate; FPP = *trans,trans*-[1-^{3}H]farnesyl pyrophosphate

enzyme might be involved in the synthesis of both C_{10} and C_{15} analogs. If this was the case, one might expect the analogous monoterpenes and sesquiterpenes to co-occur; yet, a search of several essential oil compendia [80–84] suggests such cooccurrence is uncommon. As a number of soluble enzyme preparations capable of synthesizing common monoterpenes from geranyl pyrophosphate are available, the above speculation was tested directly. The ability of these preparations to synthesize a known monoterpene from [1-^{3}H]geranyl pyrophosphate was, thus, compared to the ability of the same preparation to synthesize the analogous sesquiterpene from *trans, trans*-[1-^{3}H]farnesyl pyrophosphate (Table 1). *Trans, trans*-[1-^{3}H]farnesyl pyrophosphate was not a very efficient substrate for these monoterpene cyclase preparations, and the appropriate sesquiterpene product was detectable only in the case of monocyclic compounds. Thus, even if *trans, trans*-farnesyl pyrophosphate is available to the enzyme, it seems unlikely that the biosynthesis of sesquiterpenes by monoterpene cyclases has any significance *in vivo*.

Relatively few cell-free systems capable of sesquiterpene biosynthesis are known. Caraway (*Carum carvi*) leaves contain relatively high levels of germacrene D [85], and a 105,000 g supernatant prepared from young leaves was capable of synthesizing germacrene D from *trans, trans*-[1-^{3}H]farnesyl pyrophosphate. When [1-^{3}H]geranyl pyrophosphate was used as the substrate with this relatively crude preparation, several monoterpenes were synthesized (notably limonene), but the monoterpene analog of germacrene D (i.e., β-phellandrene, which is a component of caraway oil [86]) was barely detected (Table 1). These results provide another indication that monoterpene and sesquiterpene analogs are not synthesized by the same enzymes, and they further indicate that the distinct monoterpene synthetases and sesquiterpene synthetases are rather specific for prenyl pyrophosphate substrates of the appropriate C_{10} or C_{15} chain length. Thus, while it is quite likely that the mechanisms involved in the biosynthesis of analogous monoterpenes and sesquiterpenes are similar, comparative studies with monoterpene cyclases and the corresponding sesquiterpene cyclases will be required to explore this point further.

Concluding Remarks

From Wallach's "Isoprene Rule" to Ruzicka's "Biogenetic Isoprene Rule", the monoterpenes and sesquiterpenes have played a significant role in the development of basic concepts concerning the origin of all terpenoid compounds. Yet, while our understanding of the biosynthesis of higher terpenoids, such as steroids, has flourished, progress in the biochemistry of monoterpenes and sesquiterpenes has been relatively slow–due in large part to the chemical nature of the mono- and sesquiterpenes (volatility, instability, etc.) and to the experimental difficulties associated with the plants that produce these compounds. Since the last International Symposium, however, improvements in analytical methods and techniques for plant enzymology have lead to significant advances with both the *in vivo* and *in vitro* experimental approach. Progress with cell-free preparations, particularly in the area of monoterpene cyclases and their mechanisms of action, has been very rapid. Enzymes that interconvert the monoterpenes, and thereby increase the complexity of essential oils, are being purified and characterized. Perhaps, for the first time, the door is open to understanding monoterpene catabolism, its regulation, and its physiological rationale. The tools for comparing the origin of monoterpenes and sesquiterpenes on a molecular level are now available. Thus, recent work has opened up, more than anytime in the past, the area of monoterpene and sesquiterpene biosynthesis, and exciting developments are likely to occur in the very near future.

Acknowledgements

The work by the author cited herein was supported by grants from the U.S. National Science Foundation (PCM 76-23632 and PCM 78-19417), the Research Corporation, the Washington Mint Commission and the Mint Industry Research Council.

References

[1] Guenther, E., The Essential Oils, Vol. I, R. E. Krieger Publ. Co., New York, 1972 (Reprinted).
[2] Parry, J. W., Spices, Vol. I, Chemical Publ. Co., New York, 1969.
[3] Paech, K., Biochemie und Physiologie der Sekundären Pflanzenstoffe, Springer-Verlag, Berlin, 1950.
[4] Swain, T., Ann Rev. Plant Physiol. **28,** 479 (1977).
[5] Harborne, J. B., Introduction to Ecological Biochemistry, Academic Press, New York, 1977.
[6] Loomis, W. D., and R. Croteau, Recent Adv. Phytochem., **6,** 147 (1973).
[7] Wallach, O., Liebigs Ann. Chem. **238,** 78 (1887).
[8] Sanadze, G. A., and G. M. Dolidze, Soobshch. Akad. Nauk Gruz. SSR **27,** 747 (1961).
[9] Ruzicka, L., A. Eschenmoser and H. Heusser, Experientia **9,** 357 (1953).
[10] Beytia, E. D., and J. W. Porter, Ann. Rev. Biochem. **45,** 113 (1976).
[11] Nes, W. R., and M. L. McKean, Biochemistry of Steroids and Other Isopentenoids, University Park Press, Baltimore, 1977.
[12] Porter, J. W. (ed.), Biosynthesis of Polyisoprenoids, John Wiley & Sons, New York, in press.
[13] Valenzuela, P., E. Beytia, O. Cori and A. Yudelevich, Arch. Biochem. Biophys. **113,** 536 (1966).
[14] Loomis, W. D., in J. B. Pridham, Terpenoids in Plants, pp. 59–82. Academic Press, New York, 1967.
[15] Charlwood, B. V., and D. V. Banthorpe, Progr. in Phytochem. **5,** 65 (1978).
[16] Loomis, W. D., and R. Croteau, in P. K. Stumpf, The Biochemistry of Plants, Vol. 4, pp. 363–418. Academic Press, New York, 1980.
[17] Croteau, R., in J. W. Porter, Biosynthesis of Polyisoprenoids, John Wiley & Sons, New York, in press.
[18] Gleizes, M., Annee Biol. **15,** 101 (1976).
[19] Schütte, H. R., Progr. in Bot. **38,** 129 (1976).
[20] Rücker, G., Angew. Chem. (Int. Ed.) **12,** 793 (1973).
[21] Cordell, G. A., Chem. Rev. **76,** 425 (1976).
[22] Arigoni, D., Pure Appl. Chem. **41,** 219 (1975).
[23] Cane, D. E., in J. W. Porter, Biosynthesis of Polyisoprenoids, John Wiley & Sons, New York, in press.
[24] Chinese Materia Medica 34.17 (as cited in P. de Mayo, The Chemistry of Natural Products, Vol. II, Mono- and Sesquiterpenes (K. W. Bentley, ed.), p. 134, Interscience, New York, 1959.
[25] Finch, Jr., A. M. T. and W. R. Vaughan, J. Am. Chem. Soc. **91,** 1416 (1969).
[26] Libavius, Alchymia, J. Saurius, I. P. Kopsii, Frankfurt, 1595.
[27] Dumas, J. B., Annalen **6,** 245 (1833).
[28] Bredt, J., Ber. **26,** 3047 (1893).
[29] Bean, N. E., Chem. in Britain **8,** 386 (1972).
[30] Wagner, G., J. Russ. Phys. Chem. Soc. **31,** 680 (1899).
[31] Meerwein, H., Liebigs Ann. Chem. **376,** 152 (1910); **405,** 129 (1914); **417,** 255 (1918).
[32] Ashby, E. C., and J. R. Boone, J. Org. Chem. **41,** 2890 (1976).
[33] Lange, G. L., and J. M. Conia, Nouveau J. Chim. **9,** 311 (1977).
[34] Steele, W. V., J. Chem. Thermo. **9,** 311 (1977).
[35] Banthorpe, D. V., and D. Baxendale, J. Chem. Soc. (C), 2694 (1970).
[36] Battersby, A. R., D. G. Laing and R. Ramage, J. Chem. Soc. (Perkin I), 2743 (1972).

[37] Croteau, R., and F. Karp, Biochem. Biophys. Res. Commun. **72,** 440 (1976).
[38] Croteau, R., and F. Karp, Arch. Biochem. Biophys. **184,** 77 (1977).
[39] Croteau, R., and F. Karp, Arch. Biochem. Biophys. **198,** 523 (1979).
[40] Croteau, R., M. Felton, F. Karp and R. Kjonaas, Plant Physiol., in press.
[41] Croteau, R., C. L. Hooper and M. Felton, Arch. Biochem. Biophys. **188,** 182 (1978).
[42] Croteau, R., and M. Felton, Phytochemistry **19,** 1343 (1980).
[43] Croteau, R., and F. Karp, Arch. Biochem. Biophys. **198,** 512 (1979).
[44] Croteau, R., and F. Karp, Arch. Biochem. Biophys. **179,** 257 (1977).
[45] Cori, O., Arch. Biochem. Biophys. **135,** 416 (1969).
[46] George-Nascimento, C., and O. Cori, Phytochemistry **10,** 1803 (1971).
[47] Poulose, A. J., and R. Croteau, Arch. Biochem. Biophys. **191,** 400 (1978).
[48] Chayet, L., R. Pont-Lezica, C. George-Nascimento and O. Cori, Phytochemistry, **12,** 95 (1973).
[49] Croteau, R., M. Felton and R. C. Ronald, Arch. Biochem. Biophys. **200,** 524 (1980).
[50] Gaubius, Adversariorum varii argumenti, Liber unus, Leiden, 1771.
[51] Croteau, R., and C. Martinkus, Plant Physiol., **64,** 169 (1979).
[52] Sakata, I., and T. Mitsui, Agr. Biol. Chem. **39,** 1329 (1975).
[53] Sakata, I., and K. Koshimizu, Agr. Biol. Chem. **42,** 1959 (1978).
[54] Dumas, J. B., Annalen **6,** 152 (1833).
[55] Brühl, J. W., Ber. **21,** 457 (1888).
[56] Beckmann, E. D., Liebigs Ann. Chem. **250,** 350 (1889).
[57] Semmler, F. W., Ber. **25,** 3513 (1892).
[58] Fleischer, J., H. J. Kaminsky, R. Hopp, K. Bauer, H. Mack and M. Koepsel, Proceeds 6th Internatl. Essent. Oil Congress (paper no. 142) San Francisco (1974); see also Biedermann, W., Continuous Process for DL-Menthol, U.S. patent 4058571 (Nov. 15, 1977).
[59] Simonsen, J. L., The Terpenes, Vol. I, Cambridge University Press, London, 1931.
[60] Orloff, H. D., Chem. Rev. **54,** 366 (1954).
[61] Kremers, R. E., J. Biol. Chem. **50,** 31 (1922).
[62] Reitsema, R. H., J. Am. Pharm. Assoc. (Sci. Ed.) **47,** 267 (1958).
[63] Battaile, J., and W. D. Loomis, Biochim. Biophys. Acta **51,** 545 (1961).
[64] Burbott, A. J., and W. D. Loomis, Plant Physiol. **42,** 20 (1967).
[65] Battaile, J., A. J. Burbott and W. D. Loomis, Phytochemistry **7,** 1159 (1968).
[66] Burbott, A. J., and W. D. Loomis, Plant Physiol. **44,** 173 (1969).
[67] Burbott, A. J., and W. D. Loomis, Phytochemistry, in press.
[68] Croteau, R., and F. Karp, Arch. Biochem. Biophys. **176,** 734 (1976).
[69] Gildemeister, E., and F. Hoffmann, Die ätherischen Öle, Vol. IIIa, p. 78, 4th Ed. (W. Treibs, ed.), Akademie-Verlag, Berlin, 1960.
[70] Croteau, R., Hort Science **12,** 563 (1977).
[71] Ghosh, M. L., and S. K. Chatterjee, Ind. J. Expt. Biol. **14,** 366 (1976).
[72] Martinkus, C., and R. Croteau, Plant Physiol., in press.
[73] Croteau, R., and C. L. Hooper, Plant Physiol. **61,** 737 (1978).
[74] Herout, V., in T. W. Goodwin, Aspects of Terpenoid Chemistry and Biochemistry, pp. 53–94, Academic Press, New York, 1971.
[75] Devon, T. K., and A. I. Scott, Handbook of Naturally Occurring Compounds, Vol. II, Terpenes, Academic Press, New York, 1972.
[76] Hendrickson, J. B., Tetrahedron **7,** 82 (1959).
[77] Parker, W., J. S. Roberts and R. Ramage, Quart. Rev. Chem. Soc. **21,** 331 (1967).
[78] Andersen, N. H., Y. Ohta and D. D. Syrdal, in E. E. Van Tamelen, Bioorganic Chemistry, Vol. II, pp. 1–37, Academic Press, New York, 1978.
[79] Corbella, A., P. Gariboldi, G. Jommi and M. Sisti, J. Chem. Soc. (Chem. Commun.), 288 (1975).
[80] Guenther, E., The Essential Oils, R. E. Krieger Publ. Co., New York, 1972–1976 (Reprinted).
[81] Gildemeister, E., and F. Hoffmann, Die ätherischen Öle, 4th Edition (W. Treibs, ed.), Akademie-Verlag, Berlin, 1956–1963.
[82] Karrer, W., Konstitution und Vorkommen der organischen Pflanzenstoffe (exclusive Alkaloide), Birkhäuser Verlag, Basel, 1958.

[83] Lawrence, B. M., Progress in essential oils, No. 21, Perf. Flavorist, **4**, 50 (1979) (and preceding reports of this series).
[84] Gilbertson, G., and R. T. Koenig, Essential Oils and Related Products. Anal. Chem. **51**, 183R (1979) (and preceding reviews of this series).
[85] Lawrence, B. M., R. J. Reynolds Tobacco Co., personal communication (1979).
[86] Salveson, A., and A. Baerheim-Svendsen, Planta Med. **30**, 93 (1976).

2. SIMPLE SYNTHESES OF SESQUITERPENES

P. Weyerstahl*

Summary Terpenes can be synthesized by joining a suitable C_5- or C_{10}-unit to another such C_5- or C_{10}-unit. Some 50 years ago, L. Ruzicka made use of this strategy to obtain β-bisabolene via a diene synthesis between isoprene and myrcene. A renewed investigation of the complex reaction mixture indicated the presence of both inter- and intra-molecular addition reaction products. An alternate synthesis of mono- and sesquiterpenes was achieved by reacting the 1,3-dithiane of an isoprenoid C_5- or C_{10}-aldehyde with prenyl or geranyl bromide followed by hydrolysis or hydrogenolysis. Thus, egomaketone, perillene and dendrolasin were synthesized from furan-3-carbaldehyde. Similarly, α-turmerone and α-curcumene were obtained from prenal and 9-bromo-*p*-cymene. Furthermore, myoporone was the product of the two-fold addition reaction of an α,ω-bifunctional isoprene unit with furan-3-carbaldehyde. A new synthesis of α-bisabolol was carried out via the prenylation of 4-acetyl-1-methylcyclohex-1-ene which was the product of a regioselective Diels-Alder reaction between isoprene and methyl vinyl ketone. The mixture of isomers was separated by distillation and the appropriate C_{14}-ketone was transformed into α-bisabolol by reacting with methyl magnesium bromide.

Introduction

In the history of the synthesis of sesquiterpenes several periods can be distinguished. The first of these, starting some fifty years ago with the pioneering work of Ruzicka [1], involved the preparation of simple acyclic or monocyclic sesquiterpenes by methods generally involving many synthetic steps. The rapid development of synthetic methods in the 1960's brought about a second period in which a great number of sesquiterpenes, some with rather complicated structures, were prepared. Finally, in the last ten years, at least four directions of sesquiterpene synthetic research can be discerned: 1, improved syntheses of major sesquiterpenes which allow higher yields; 2, syntheses with stereoselective steps thus avoiding complicated separations; 3, enantioselective syntheses starting either from a chiral pool or with the aid of enantioselective reagents; and 4, simple syntheses of sesquiterpenes of potential use on an industrial scale. Examples of these four approaches to sesquiterpene synthesis can be found in a number of recent review articles [2].

* P. Weyerstahl, Department of Organic Chemistry, Technical University Berlin, 1000 Berlin, West Germany.

Over the last few years we have examined some problems of simple synthesis of sesquiterpenes involving considerations of both starting materials and synthetic methods. Our choice of starting materials was restricted to naturally occurring or industrially available C_5- or C_{10}- units as building blocks, and, as preferred synthetic methods, we investigated Diels-Alder reactions, Claisen rearrangements and alkylations.

Diels-Alder Reaction of Isoprene and Myrcene

The ideal simple synthesis combines a C_5- and a C_{10}-compound in one step, with quantitative formation of a naturally occurring sesquiterpene. While Ruzicka had first attempted such a synthesis nearly fifty years ago, it is important to state at the outset that this goal has not yet been reached. Ruzicka investigated the reaction of isoprene with myrcene at 225 °C in sealed tubes, from which he expected four reaction products [3] (Fig. 1). As one

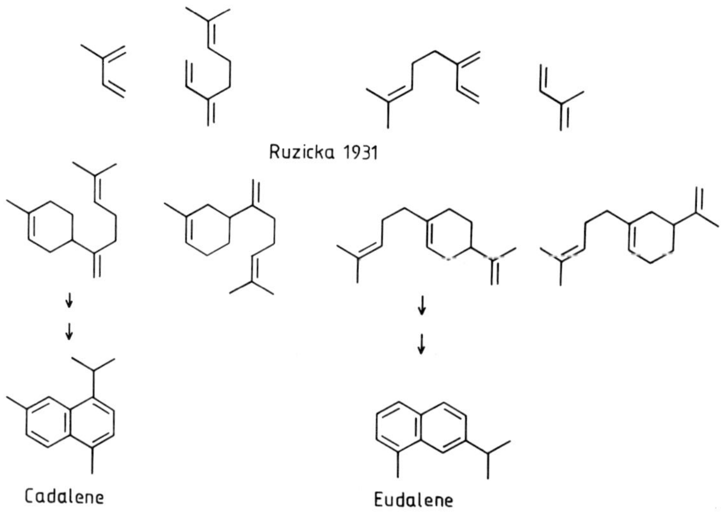

Fig. 1 The reaction of isoprene with myrcene was thought by Ruzicka [3] to yield four isomers, two of which (the 1,4-isomers) would give rise to cadalene and eudalene.

possibility, isoprene can react as a diene and myrcene as a dienophile at the least substituted double bond, yielding two isomeric products (i.e., the 1,4- and 1,5-isomers). As a second possibility, myrcene can react as the diene and isoprene as a dienophile at the least substituted double bond, and again both possible isomers are expected. Cyclization and aromatization of the 1,4-isomers should then lead to well known products, either cadalene or eudalene. However, contrary to expectation, Ruzicka found eudalene only after the four step reaction sequence outlined in Figure 2. Ruzicka treated the reaction products with hydrochloric acid, isolated a solid chlorohydrate, and assigned the structure indicated (Fig. 2). Elimination with an acetic acid/sodium acetate mixture allowed recovery of the hydrocarbon, which was cyclized by boiling with formic acid to a compound for which Ruzicka proposed the indicated structure. Finally, heating at 320°C with pulverized selenium yielded eudalene. These results forced Ruzicka to assign the given hydrocarbon structure (Fig. 2) for the Diels-Alder product of isoprene with myrcene, despite the fact that all of these reactions occurred with low yield.

We repeated this fascinating reaction of isoprene with myrcene, and investigated the reaction products as completely as possible using modern equipment and techniques, including effective distillation columns, gas chromatography and NMR spectroscopy [4]. In a preliminary distillation of the reaction products, four fractions containing C_5-, C_{10}-, C_{15}- and C_{20}-hydrocarbons were obtained. Higher homologs ($>C_{20}$) and the C_5-fraction were not examined.

The C_{10}-fraction contained two groups of hydrocarbons, as expected from the dimerization of isoprene in the Diels-Alder reaction. All four of the possible products were observed, with a ratio of 2 : 1 for the 1,4- and the 1,5-isomers (Fig. 3). A ratio of about 4 : 1 was noted regarding the

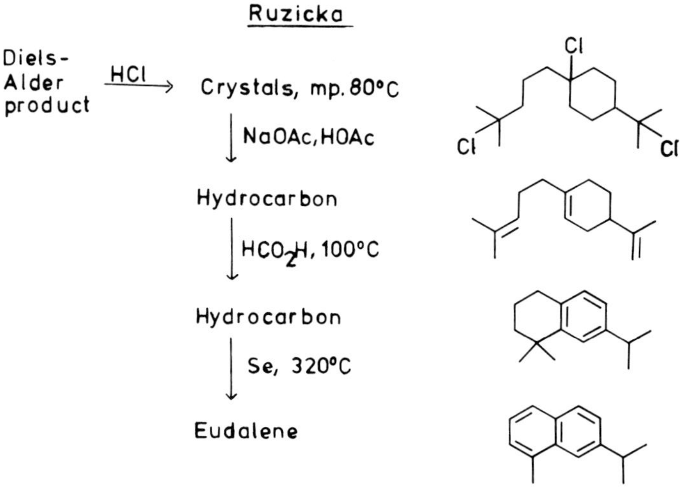

Fig. 2 Four step reaction sequence used by Ruzicka [3] to prepare eudalene.

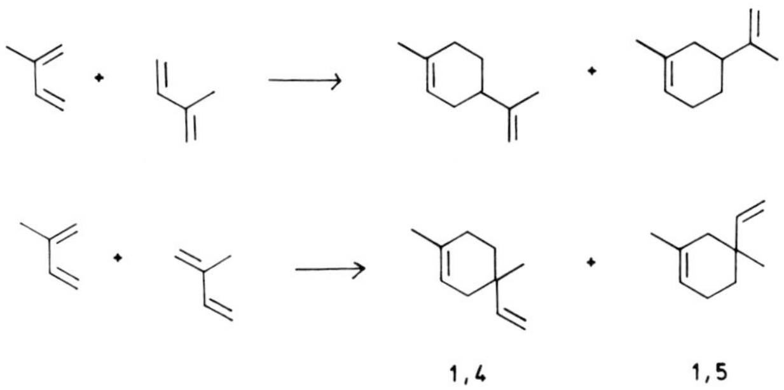

Fig. 3 Hydrocarbons obtained from the dimerization of isoprene.

preference for the less substituted double bond of the dienophile. Besides these products, myrcene (starting material) and some of its rearrangement products were found in minor quantities.

The dimerization products of myrcene were found in the C_{20}-hydrocarbon fraction, and the main component was α-camphorene which is a natural product (Fig. 4). Comparison of product ratios again showed about a 2:1 preference for 1,4-addition over 1,5-addition, and reaction of the monosubstituted double bond was again favored (about three-fold).

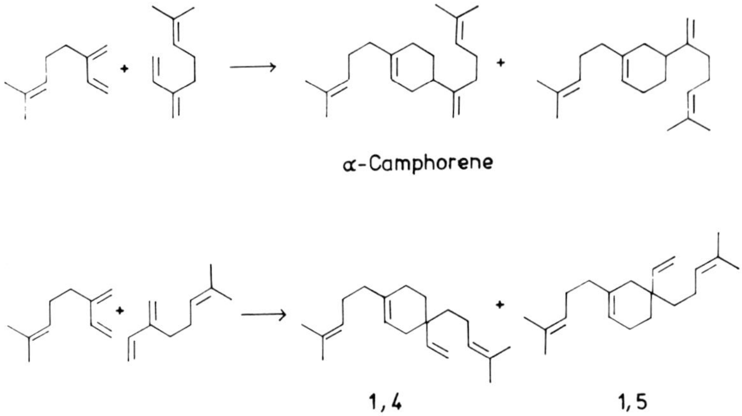

Fig. 4 Hydrocarbons obtained from the dimerization of myrcene.

The C_{15}-hydrocarbon fraction was the most interesting, containing five pairs of 1,4- and 1,5-isomers (Fig. 5). The products on the top line (Fig. 5) result from myrcene acting as a diene and isoprene as a dienophile. Ruzicka's hydrocarbon [3] is at the top left corner. The second line shows the products formed with isoprene as the diene and myrcene as the dienophile. It should be noted here that the trisubstituted double bond does not react to a significant degree. Finally, the bottom line illustrates two cyclooctadienes, products of a Ziegler-type 4+4-cycloaddition. The C_{15}-fraction was separated on a one kg scale by careful distillation on an automatic two meter column. In the case of minor products, the isomers remained unresolved, and these are separable only by gas chromatography. The structural assignments of all of these compounds were determined by a combination of high resolution proton- and ^{13}C-NMR spectroscopy. Additionally, these products were compared with authentic bisabolene and with other authentic standards synthesized in an unambiguous way. Ruzicka's hydrocarbon was the main component (44 %) of the C_{15}-fraction, while the isomeric compound (Fig. 5, top row second from left) was formed in 22 % yield. β-bisabolene, on the other hand, was obtained in only 8 % yield after separation by distillation. Catalysis of Diels-Alder reactions with Lewis acids is known to effect regioselectivity favoring the 1,4-products [5]. Thus, the way is now open for catalyzing the reaction to β-bisabolene as the major product.

Reaction of Ruzicka's hydrocarbon and its 1,5-isomer with HCl affords a pair of chlorohydrates (Fig. 6). Both of these compounds have the same melting point, within expected

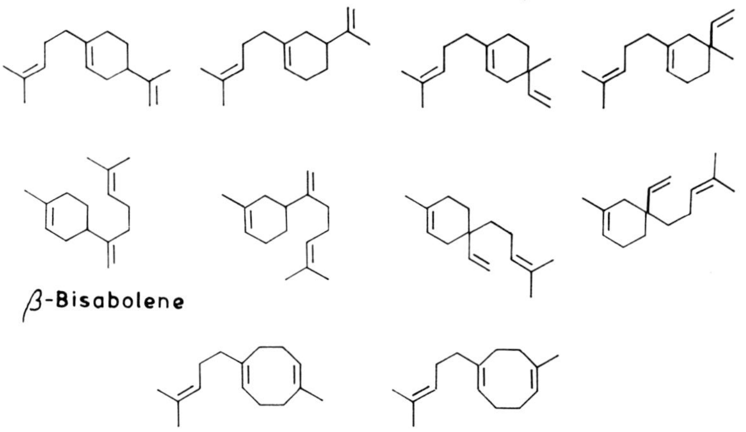

Fig. 5 Components of the C_{15}-hydrocarbon fraction obtained from the reaction of isoprene with myrcene.

error, and the same IR and mass spectra [4]. Even the 270 MHz proton-NMR spectra were identical. Thus, the addition was clearly according to Markownikov's Rule, yet there was no indication that the two compounds were different from one another. However, the ^{13}C-NMR spectra did show different signals, which were readily detectable even in the chlorohydrate mixture prepared according to Ruzicka [3], illustrating the power of ^{13}C-NMR as an important analytical tool. The mixture of chlorohydrates exhibits a melting point of 80 °C which cannot be increased by recrystallization.

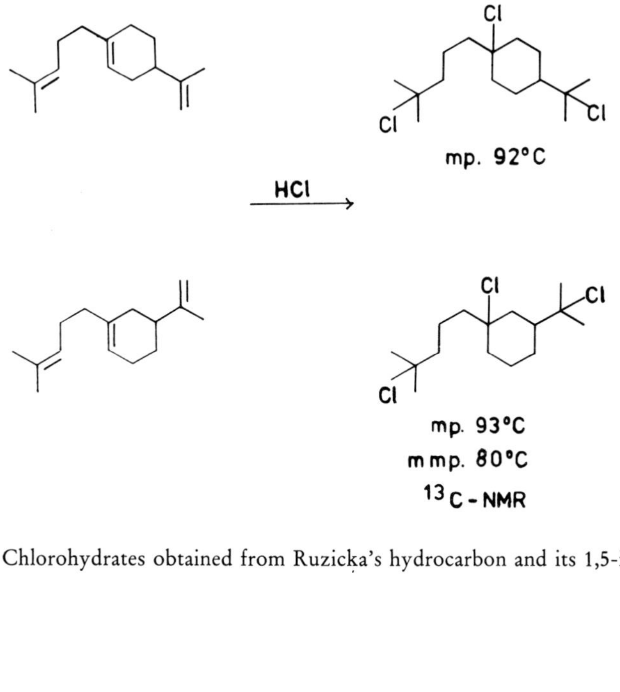

Fig. 6 Chlorohydrates obtained from Ruzicka's hydrocarbon and its 1,5-isomer.

Application of the Diels-Alder Reaction to the Synthesis of α-Bisabolene and β-Santalenone

Figure 7 illustrates two isomeric bisabolenes, *E*-α-bisabolene and *E*-*iso*-α-bisabolene. Both compounds occur naturally, and they were synthesized by Vig and associates [6] starting from the C_{14}-ketone indicated (Fig. 7). These synthetic steps are well known, and the yields are good. Clearly, the key compound of the reaction sequence is the C_{14}-ketone which, along with its *Z*-isomer, was synthesized by Vig and associates [7] by a multistep procedure. Application of the Diels-Alder reaction allowed this synthesis to be greatly simplified [8]. Thus, Diels-Alder

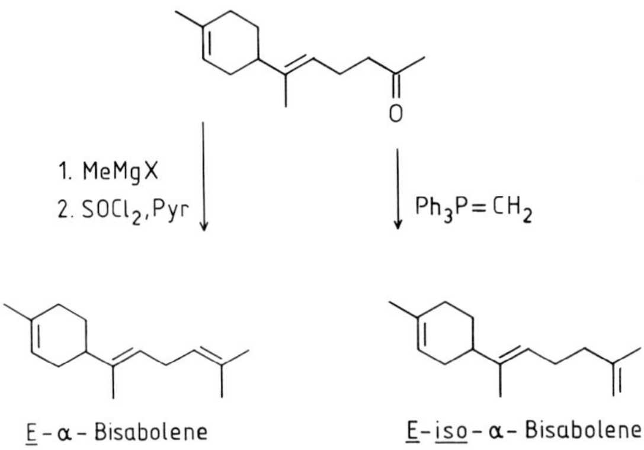

Fig. 7 Preparation of *E*-α-bisabolene and *E*-*iso*-α-bisabolene from the C_{14}-ketone according to the procedure of Vig [6].

reaction of isoprene with methyl vinyl ketone catalyzed by $SnCl_4$ resulted in the 1,4-type cyclohexenyl ketone in 85 % yield and 97 % purity (Fig. 8). When necessary, even higher purity can be readily obtained by fractional distillation. This very useful C_9-ketone, discussed again later, was reacted with vinyl magnesium halide to afford the allylic alcohol (Fig. 8). For the production of the alcohol on a technical scale, the reaction of the ketone with acetylene, followed by selective hydrogenation, would probably be preferred. The tertiary allylic alcohol was then heated in a pressure vessel with isopropenyl ether in the presence of phosphoric acid yielding, in one step, the thermodynamically stable mixture of the *E*- and *Z*-C_{14}-ketones. Although a C_5-building block was used only once, this three-step synthesis provides a simple route to the *E*-isomer, which can be separated from the equilibrium mixture by fractional distillation.

Less simple, but probably more stimulating, is the synthesis of β-santalenone [9]. Starting again with a Diels-Alder synthesis using the C_5-unit cyclopentadiene, the norbornadiene derivative indicated (Fig. 9) was obtained in good yield (78 %). Selective hydrogenation of the less substituted double bond and reduction of the ester to the corresponding allylic alcohol are very simple procedures, and the one-step Claisen rearrangement, carried out by heating this alcohol with the enol ether in the presence of mercury acetate, readily gave β-santalenone (68 %

yield). This synthesis, described in detail elsewhere [10], seems very simple, but it does require the dienol ether as a crucial synthon. The synthesis of this dienol ether, as described in the literature [11], is not easy. However, this compound is the enol ether of mesityloxide, a readily available material, and a direct route might be found.

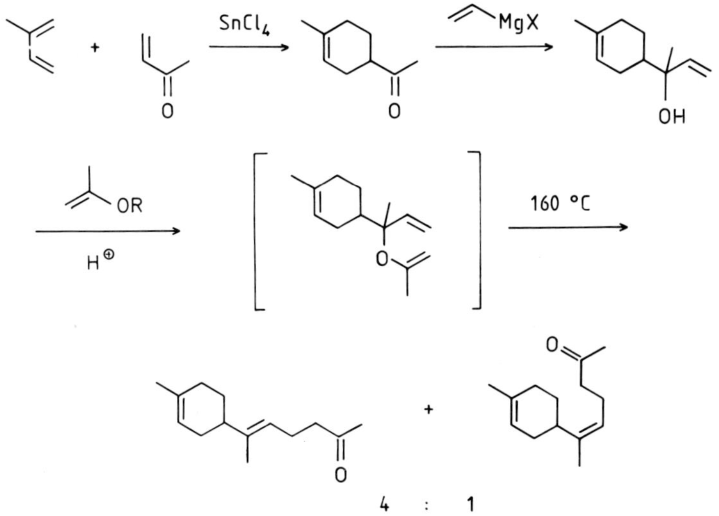

Fig. 8 Application of the Diels-Alder reaction and the Claisen rearrangement to the synthesis of the C_{14}-ketone.

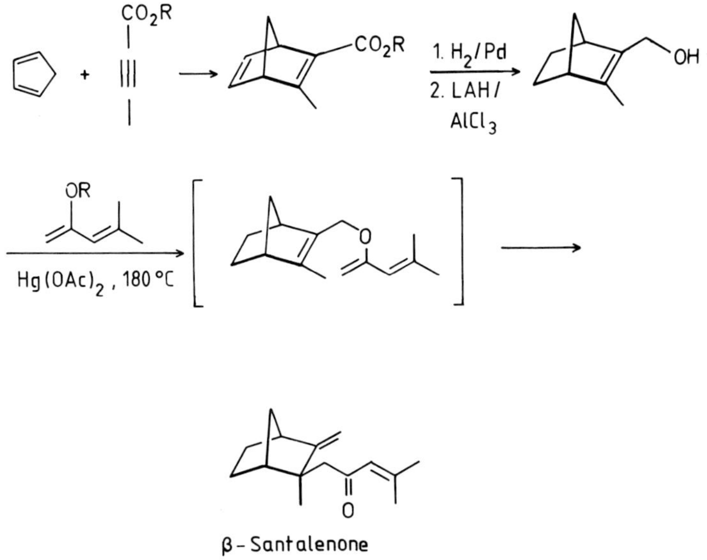

Fig. 9 Application of the Diels-Alder reaction and the Claisen rearrangement to the synthesis of β-santalenone.

Application of Dithianes to the Synthesis of Sesquiterpenes

The well known reaction of an aldehyde and 1,3-propane dithiol, in the presence of a Lewis acid, readily leads to the corresponding substituted 1,3-dithiane in nearly quantitative yield (Fig. 10). Due to the acidity of the proton at C-2, deprotonation with a strong base gives the anion or the lithium salt, respectively, which can be alkylated by any alkyl halide to the corresponding disubstituted dithiane [12] (Fig. 10). Thus, this reaction, a so-called "Umpolung", leads from an aldehyde derivative to the respective ketone derivative. The ketone can be easily liberated by mild hydrolysis in the presence of mercury salts. Additionally, the dithiane can be hydrogenated by treating with Raney nickel to give the saturated hydrocarbon. As applied to our sesquiterpene synthesis program, we therefore chose a C_5- or C_{10}-aldehyde to give the respective dithiane and then alkylated with a C_{10}- or C_5-halide. Thus, this approach allows the synthesis of a great variety of C_{15}-ketones and hydrocarbons.

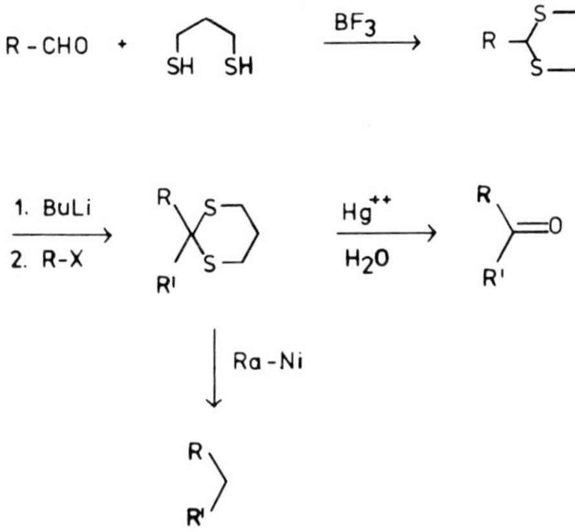

Fig. 10 Preparation and some reactions of substituted 1,3-dithianes.

The first line of Figure 11 illustrates three simple C_{10}-aldehydes, citral, myrtenal, and perilla aldehyde. All these aldehydes are α,β-unsaturated, and are therefore much more reactive than saturated aldehydes. Although some difficulties with the Corey-Seebach reaction were expected, in all cases the first step, as well as the deprotonation, proceeded smoothly [13]. The alkylation reagent chosen was prenyl bromide, a second very reactive compound; yet, the doubly-unsaturated dithiane was readily formed. Thus, the crucial steps were the hydrolysis and, particularly, the hydrogenolysis. Although the reaction conditions had to be very carefully maintained, six sesquiterpenes were synthesized [13] (Fig. 12). The hydrocarbon at the upper right hand corner (Fig. 12) is identical to Ruzicka's hydrocarbon (Fig. 2).

A special type of α,β-unsaturated aldehyde is represented by furan-3-carbaldehyde (Fig. 13). The corresponding dithiane is readily formed, but this compound has two acidic protons, the dithiane proton and the α-proton. Fortunately, the dithiane proton is the more acidic, and this anion was alkylated with prenyl bromide to give perillene after hydrogenolysis, and egomaketone after hydrolysis [13] (Fig. 13). Similarly, alkylation with geranyl bromide

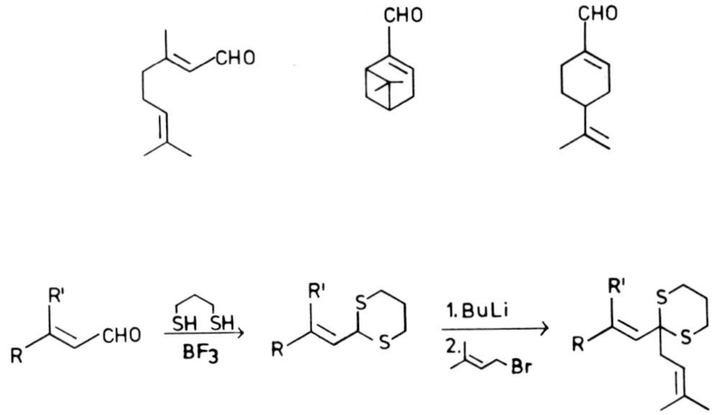

Fig. 11 Structures of citral, myrtenal and perilla aldehyde, and preparation of the corresponding prenylated terpenyl dithianes.

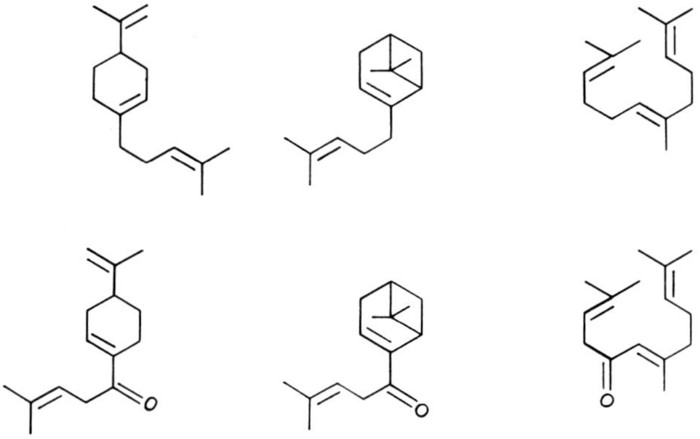

Fig. 12 Sesquiterpene hydrocarbons and ketones prepared from the corresponding dithianes of Fig. 11.

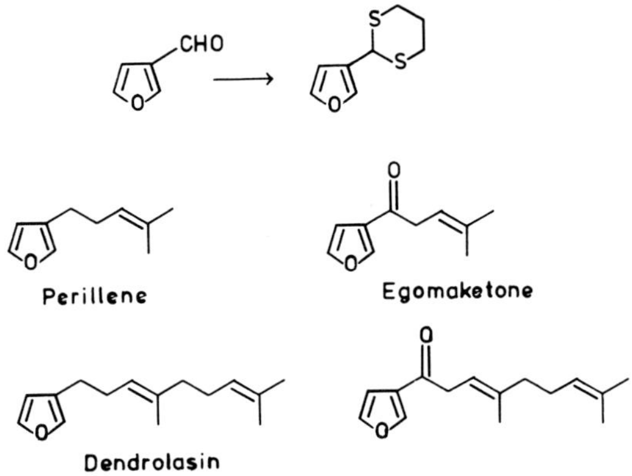

Fig. 13 The dithiane prepared from furan-3-carbaldehyde, and the products obtained via alkylation of this intermediate with prenyl bromide and geranyl bromide.

provided a dithiane which was hydrogenolyzed to dendrolasin and hydrolyzed to the corresponding ketone (Fig. 13), thus indicating the versatility of this reaction.

Figure 14 illustrates the reaction sequence from a C_{10}-aldehyde (9-oxo-*p*-cymene) which is not naturally occurring. The known 9-oxo-*p*-cymene [14] was readily obtained in an improved synthesis from commercially available *p*-methyl acetophenone via a Wittig procedure.

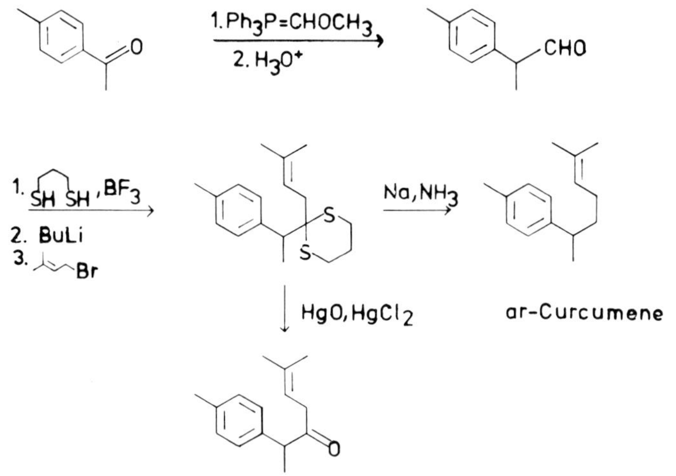

Fig. 14 Application of the dithiane reaction to the preparation of ar-Curcumene from 9-oxo-*p*-cymene.

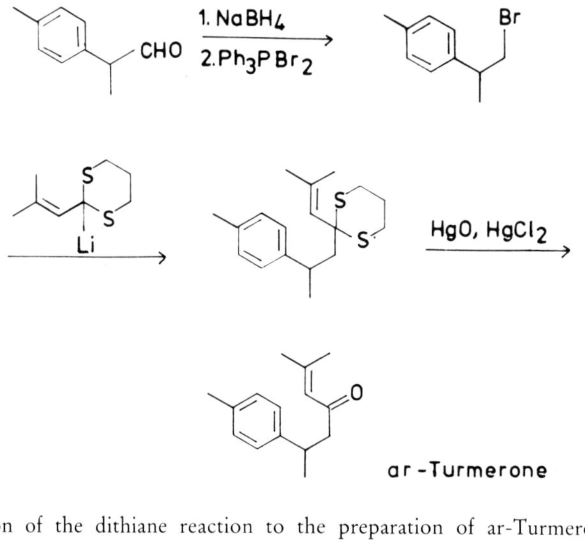

Fig. 15 Application of the dithiane reaction to the preparation of ar-Turmerone from 9-bromo-*p*-cymene.

Formation of the dithiane followed by prenylation and hydrolysis yields the unknown ketone, whereas reaction with sodium in liquid ammonia affords ar-curcumene (Fig. 14). Reduction of 9-oxo-*p*-cymene with sodium borohydride, and subsequent bromination, provided a new C_{10}-bromide which was reacted with the C_5-dithiane derived from prenal (Fig. 15). Hydrolysis of this disubstituted dithiane in the presence of mercury oxide and mercury chloride gave ar-turmerone [15], a further example of a well-known sesquiterpene.

The syntheses described above have been applications of the C_5-unit monofunctionalized as prenyl bromide or prenal. Figure 16 illustrates the preparation of a difunctionalized isoprene useful for two further types of syntheses. Treatment of isoprene with *N*-bromosuccinimide in acetic acid solution furnished a mixture of bromoacetates in very good yield. The 1,2-addition product is a minor compound, and it can be readily removed by fractional distillation. The *E,Z*-mixture of 1,4-addition products can not be separated directly, but the *trans*-compound, the main product, can be obtained as the bromoalcohol after saponification with potassium carbonate and subsequent treatment with sodium hydride (Fig. 16). Under these conditions, the *cis*-bromoalcohol is cyclized to the dihydrofuran, while the *trans*-bromoalcohol remains. These products are readily separated. However, if a definite stereochemistry at the double bond is not required, further steps can be conveniently carried out with the mixture of isomers.

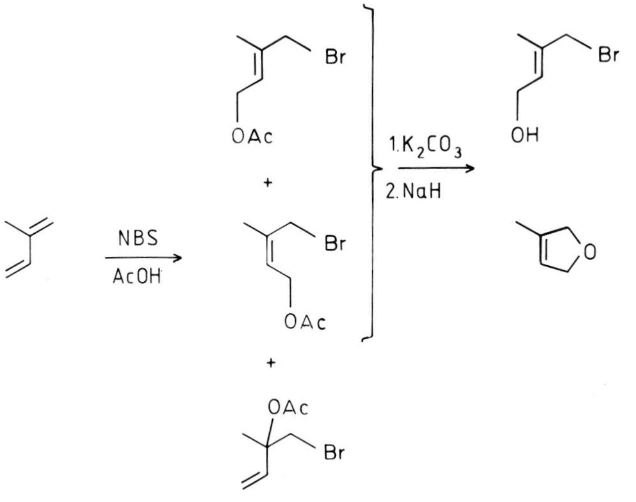

Fig. 16 Preparation of a difunctionalized isoprene unit.

Application of this difunctionalized isoprene unit is illustrated by the synthesis of the corresponding functionalized sesquiterpenes (Fig. 17). Thus, citral dithiane, when deprotonated with butyl lithium and alkylated with the protected C_5-bromoalcohol, led to a disubstituted dithiane which was cleaved by three methods [16]. Hydrolysis under the normal conditions gave the indicated hydroxyketone, a ketofarnesol, while hydrogenolysis with Raney nickel yielded farnesol (Fig. 17). Reaction of the dithiane with sodium in liquid ammonia additionally cleaved the allylic alcohol function to give the expected triene.

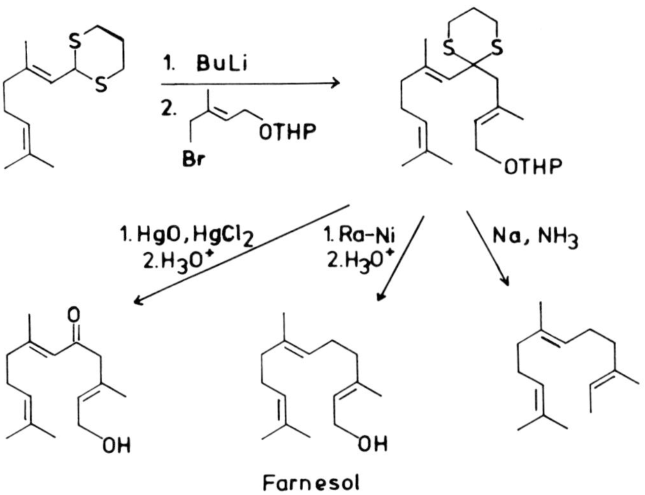

Fig. 17 Utilization of the difunctionalized isoprene unit in the synthesis of farnesane-type sesquiterpenes.

Figure 18 shows a second application of the isoprene synthon, in which both functional groups were utilized in a stepwise fashion. The dithiane of the isoprenoid isovaleraldehyde was reacted with the isomer mixture of the protected bromoalcohol to give the C_{10}-skeleton, functionalized at two positions. Hydrolysis, without mercury salts, and bromination led to the

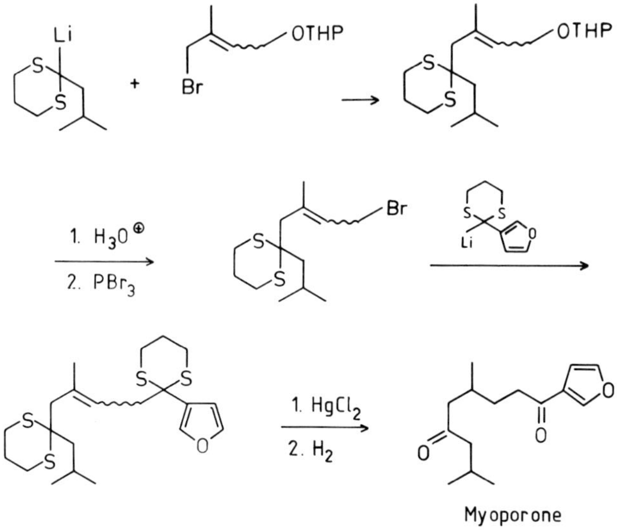

Fig. 18 Utilization of the difunctionalized isoprene unit in a stepwise manner for the synthesis of myoporone.

allylic bromide excellently suitable for an alkylation reaction (Fig. 18). Alkylation of the previously described dithiane of furan-3-carbaldehyde provided the unusual bis-dithiane. Mercury assisted hydrolysis, and subsequent hydrogenation of the double bond in the presence of homogeneous catalysts, finally yielded myoporone, a sesquiterpene diketone [17].

A Simple Synthesis of α-Bisabolol

The last step of a well-known synthesis of α-bisabolol [18] is shown in Figure 19. An essential reactant is the C_9-ketone, the reaction product of isoprene with methyl vinyl ketone described earlier. Grignard reaction with the C_6-unit shown (Fig. 19) gives α-bisabolol directly. Although not readily apparent, the great disadvantage of this synthesis is the multistep synthesis of the C_6-halide which is both expensive and difficult. A synthesis of α-bisabolol utilizing an

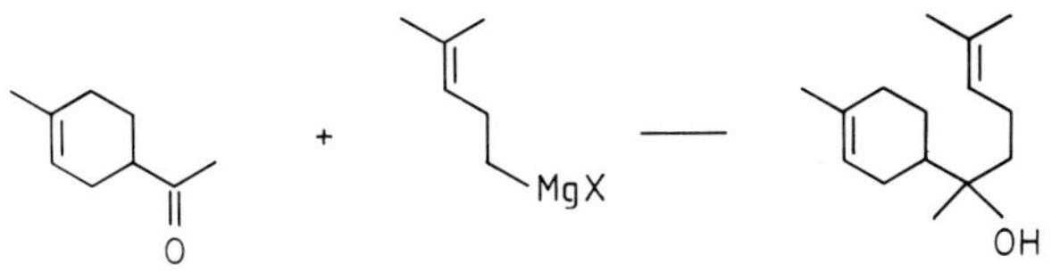

Fig. 19 Synthesis of α-bisabolol by the procedure of Ruzicka [18].

alkylation reaction with inexpensive prenyl chloride was therefore developed. Reaction of isoprene with technical grade aqueous hydrochloric acid led to a mixture of 1,4-product with 10% or less 1,2-product (Fig. 20). For alkylation reactions it is not necessary to purify the product, because even the 1,2-addition product, via allylic rearrangement, leads to the normal prenylated compound. Reaction of the ketone with prenyl chloride, in the presence of a base, provided as products the desired straight-chain C_{14}-ketone, the undesired branched-ketone, and the diprenylated compound (Fig. 21). More sophisticated bases such as sodium hydride or potassium *t*-butoxide in tetrahydrofuran were initially employed, but the reaction works well

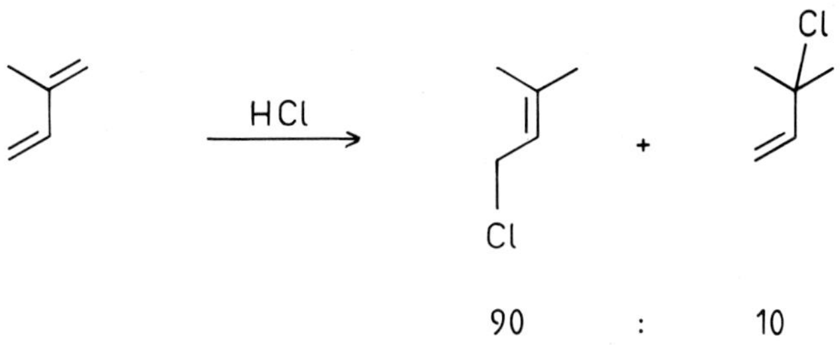

Fig. 20 Preparation of prenyl chloride.

with potassium hydroxide at 40 °C. The amount of the diprenylated product (Fig. 21) can be drastically decreased when an excess of starting ketone is employed, and any unreacted C_9-ketone can be easily recovered by distillation. The remaining difficulty was the ratio of unbranched and branched C_{14}-ketone, and this problem was approached by systematically varying all reaction parameters to optimize total yield and isomer ratio. The desired compound is, by far, the major reaction product, but there is still room for some improvement. The last step

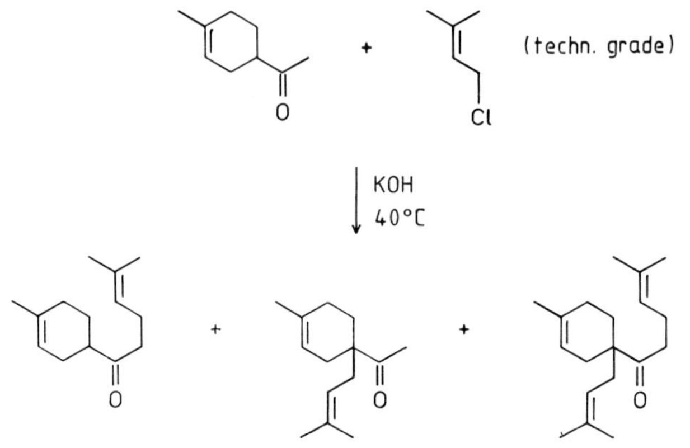

Fig. 21 Prenylation of 4-acetyl-1-methylcyclohex-1-ene.

of the bisabolol synthesis, a simple Grignard reaction with methyl magnesium halide, is well-known and carried out without difficulty (Fig. 22). Thus, the total synthesis of α-bisabolol consists of three simple steps (a Diels-Alder reaction, an alkylation, and a Grignard reaction) and four building blocks (C_5-isoprene, C_4-methyl vinyl ketone, C_5-isoprene via prenyl chloride, and C_1-methyl chloride).

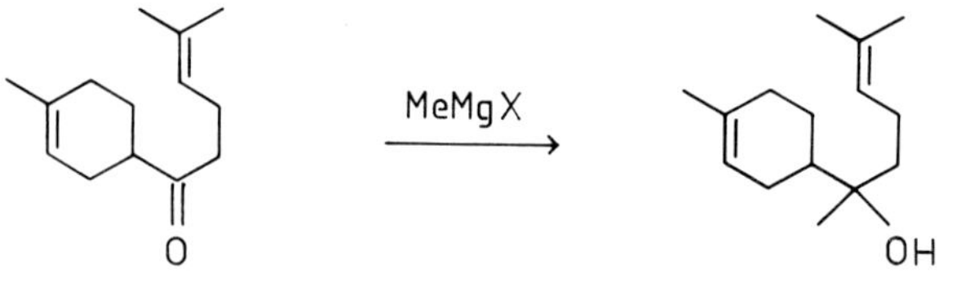

Fig. 22 Treatment of the C_{14}-ketone with methyl magnesium halide affords α-bisabolol.

Concluding Remarks

Sesquiterpenes are widely used as fragrances, and, in some instances, a simple synthesis of a sesquiterpene can mean a simple synthesis of a fragrance. However, in most cases it is a long way from the so-called "simple synthesis" in the laboratory to a process on an industrial scale.

References

[1] Ruzicka, L., Helv. Chim. Acta **6,** 492 (1923).
[2] Terpenoids and Steroids, Vol. 1–9, The Chemical Society, London, 1971–1979.
[3] Ruzicka, L., and W. Bosch, Helv. Chim. Acta **14,** 1336 (1931).
[4] Eisfelder, W., and P. Weyerstahl, Liebigs Ann. Chem. **1977,** 988.
[5] Lutz, E. F., and G. M. Bailey, J. Am. Chem. Soc. **86,** 3899 (1964).
[6] Vig, O. P., J. Chander and B. Ram, Indian J. Chem. **12,** 1156 (1974).
[7] Vig, O. P., J. C. Kapur, S. D. Sharma and B. Ram, Indian J. Chem. **11,** 1247 (1973).
[8] Becker, M., and P. Weyerstahl, Helv. Chim. Acta **62,** 2724 (1979).
[9] Simmross, F.-M., and P. Weyerstahl, Liebigs Ann. Chem., 1981, in press.
[10] Simmross, F.-M., Dissertation, Technische Universität Berlin, 1980.
[11] Normant, H., and G. Martin, Bull. Soc. Chim. Fr. **1957,** 429.
[12] Seebach, D., Synthesis **1969,** 17.
[13] Hoppmann, A., and P. Weyerstahl, Tetrahedron **34,** 1723 (1978).
[14] Johnson, W. S., J. S. Belew, L. J. Chinn and R. H. Hunt, J. Am. Chem. Soc. **75,** 4995 (1953).
[15] Bokel, H. H., A. Hoppmann and P. Weyerstahl, Tetrahedron **36,** 651 (1980).
[16] Entz, U., Diplomarbeit, Technische Universität Berlin, 1977.
[17] Entz, U., K. Meÿen and P. Weyerstahl, unpublished results.
[18] Ruzicka, L., and M. Liguori, Helv. Chim. Acta **15,** 3 (1932).

3. A REVIEW OF THE REACTION BETWEEN ALDEHYDES AND PHENOLS OR PHENOL ETHERS, AND ITS APPLICATION TO THE SYNTHESIS OF FRAGRANCE AND FLAVOR SUBSTANCES

K. A. Bauer*, A. Krempel*, R. G. Mölleken*, H. Fiege** and K. Wedemeyer**

Summary In certain cases phenolic alcohols can be obtained by the reaction between phenols and aliphatic aldehydes under alkaline conditions. These phenolic alcohols serve as the starting material for the synthesis of phenolic aldehydes. If the condensation reaction is carried out in the presence of an acidic catalyst, dihydroxydiphenylalkanes are formed which, when cleaved under *alkaline* conditions, yield the corresponding alkenylphenols and the starting phenol, which can be recycled. The reaction between phenol ethers and aldehydes performed under acidic conditions results in the corresponding alkoxydiphenylalkanes. In certain cases, the reaction stops at the carbinol stage. Alkoxydiphenylalkanes can be cleaved under *acidic* conditions to give alkenylphenyl ethers and the original phenol ether. The aforementioned investigations on the reaction between aldehydes and phenols, or phenol ethers, were carried out by Haarmann & Reimer GmbH and Bayer AG over the last few years, with the aim of improving existing processes and discovering new areas of application in the industrial synthesis of fragrance and flavor substances.

Introduction

The reaction between phenols and aldehydes is certainly one of the most thoroughly investigated reactions in the field of organic chemistry. This is especially true for the condensation reaction between phenol and formaldehyde, which gives rise to macromolecular products of great technological importance [1]. The research work reviewed herein is restricted to the industrial feasibility of a few, well-known reactions between phenols and aldehydes which could serve as the basis for the synthesis of fragrance and flavor substances. As is often the case with reactions that can give rise to polymeric products, special reaction conditions are necessary if the low molecular weight products are to be obtained. These conditions, which can be adjusted in order to steer the reaction in the desired direction, are discussed first, followed by

* K. A. Bauer, A. Krempel and R. G. Mölleken, Haarmann & Reimer GmbH, 3450 Holzminden, West Germany.
** H. Fiege and K. Wedemeyer, Bayer AG, 5090 Leverkusen, West Germany.

a description of the progress made in the reaction between phenol ethers and aldehydes. Finally, the application of the various reaction products to the synthesis of fragrance and flavor substances is reviewed.

Synthesis of Mononuclear Phenolic Alcohols and Binuclear Bis(Hydroxyphenyl)Alkanes from Phenols and Aliphatic Aldehydes

If the macromolecular products are ignored, then only two types of compounds can be obtained in good yields from phenols and aliphatic aldehydes. Fortunately, both of these are suitable intermediates for the synthesis of fragrance and flavor substances. The two types of products are mononuclear phenolic alcohols, which are products of addition reactions, and binuclear dihydroxy diaryl alkanes, which arise as a result of condensation reactions. Examples of addition reactions, in which very little by-product formation is evident, are shown in Figure 1. Likewise, typical condensation reactions, which terminate at the bis(hydroxyphenyl)alkane stage, are illustrated in Figure 2.

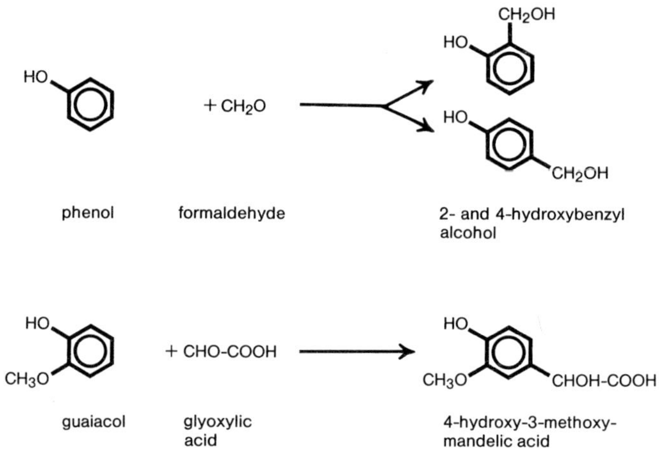

Fig. 1 Examples of addition reactions between aldehydes and phenols in which by-product formation is minimal.

The reaction between phenols and aldehydes to yield the above mentioned intermediates is catalyzed by either acid or alkali. In certain cases, homogeneous catalysis under alkaline conditions is preferred to the heterogeneous acidic alternative which utilizes, for example, ion exchange resins [2]. An example of the latter is the synthesis of bis(hydroxyphenyl)alkanes [3], whereas the former procedure is used when aliphatic aldehydes such as formaldehyde, chloral or glyoxylic acid are used for the synthesis of those carbinols which are stable and which do not react further [4]. To discuss in detail the effect of individual parameters upon the kinetics and mechanism of the reaction is not feasible here. Yet, it is important to note that the overall course

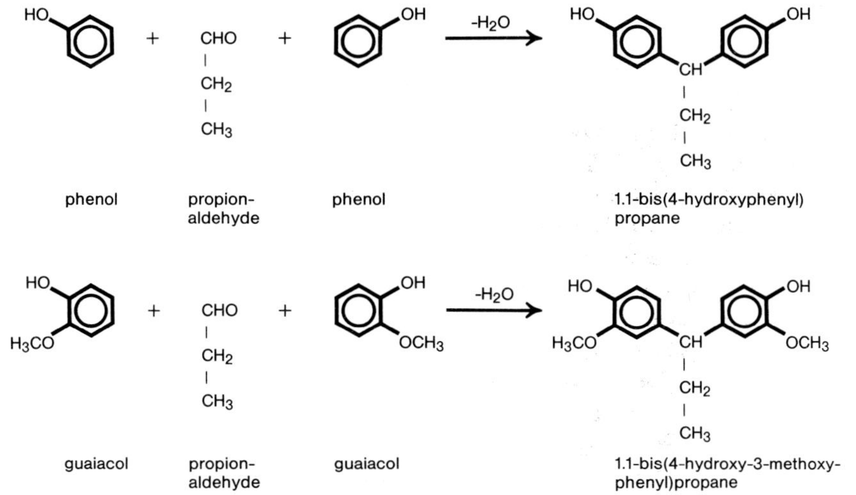

Fig. 2 Examples of condensation reactions between aldehydes and phenols.

of the reaction is influenced, to a large extent, by the nature of the catalyst employed which, thus, determines whether hemiacetals, dihydroxyalkyl phenols or binuclear compounds are preferentially formed.

While the complexities of the reaction are described in detail in paper No. 4 of this symposium, it is worth mentioning here one particular problem that confronted us, the distribution and properties of *ortho*- and *para*-substituted compounds. As indicated in Figure 1, the alkali catalyzed reaction of phenol with formaldehyde normally gives a mixture of salicyl alcohol and 4-hydroxybenzyl alcohol. As the presence of certain metal ions favors the *ortho*-mode of substitution, presumably via the formation of an intermediate chelate, a procedure for the synthesis of saligenin is known (Fig. 3, for further details see paper No. 4). Unfortunately, no catalyst has yet been found that favors the formation of the *para*-substituted derivative under similar reaction conditions. It has been shown, however, that the addition of

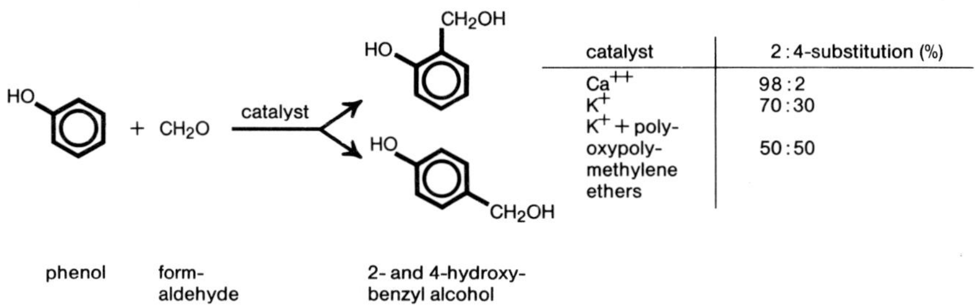

Fig. 3 Procedure for the synthesis of 2- and 4-hydroxybenzyl alcohol. For detailed description see Symposium paper no. 4.

small amounts of polyoxypolymethylene ethers to a potassium hydroxide-catalyzed reaction shifts the *para:ortho* ratio from approximately 30:70 to approximately 50:50 [5]. This allows a much greater degree of freedom in designing a procedure for the synthesis of hydroxybenzyl alcohols, at least as far as the distribution of isomers is concerned.

The isomeric phenolic alcohols are extremely reactive substances and, consequently, their separation from the complex reaction system proved to be very difficult indeed. It became obvious that a new and mild isolation procedure had to be developed if an economical synthesis of the required products was to be achieved. This procedure, which consists of a number of successive extraction steps, was developed by Alfred Krempel in our laboratories and it is described in paper No. 4 of this symposium. Fortunately, the addition of chloral or glyoxylic acid to phenols yields mainly the *para*-derivative when carried out under alkaline conditions. Consequently, the otherwise difficult removal of the *ortho* isomers, which are present in small amounts and are not wanted, can be dispensed with in this case. *Para*-substituted compounds, such as 1,1-bis(4-hydroxyphenyl)propane and 1,1-bis(4-hydroxy-3-methoxyphenyl)propane, can also be obtained in good yield by the condensation between phenols and propionaldehyde, when the reaction is carried out under strongly acidic conditions (Fig. 2) [6]. Such *para*-substituted compounds are normally of great interest, especially for the subsequent synthesis of fragrance and flavor substances. As indicated in Figure 4, the ratio of the isomers of bis(hydroxyphenyl)propane, obtained from the reaction between guaiacol and propionaldehyde, amounts to some 85:14:1.

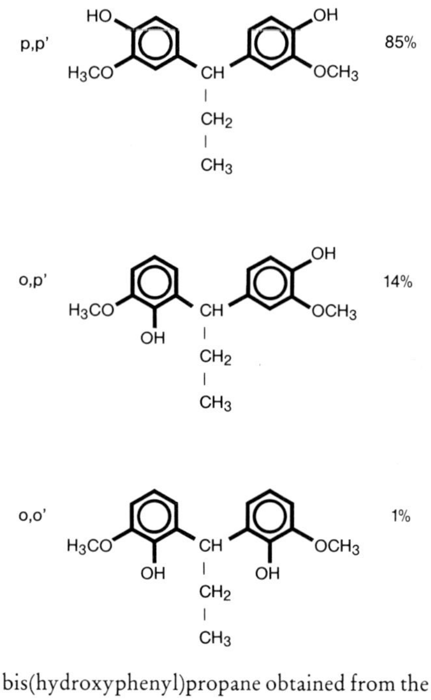

Fig. 4 Isomer rations of the bis(hydroxyphenyl)propane obtained from the reaction between guaiacol and propionaldehyde.

Synthesis of Mononuclear Phenol Ether Alcohols and Binuclear Bis(Alkoxyphenyl)Alkanes from Phenol Ethers and Aliphatic Aldehydes

The addition of aldehydes to phenols under basic reaction conditions involves the formation of a phenolate anion, with resulting activation of the phenol. To convert phenol ethers to their corresponding mono- or binuclear derivatives, highly specific acidic reaction conditions had to be developed. Such conditions were, in fact, successfully obtained in two cases of particular interest. If a mixture of glyoxylic acid and methylenedioxybenzene is warmed, bis(3,4-methylenedioxyphenyl)acetic acid is formed via autocatalysis by the glyoxylic acid (Fig. 5).

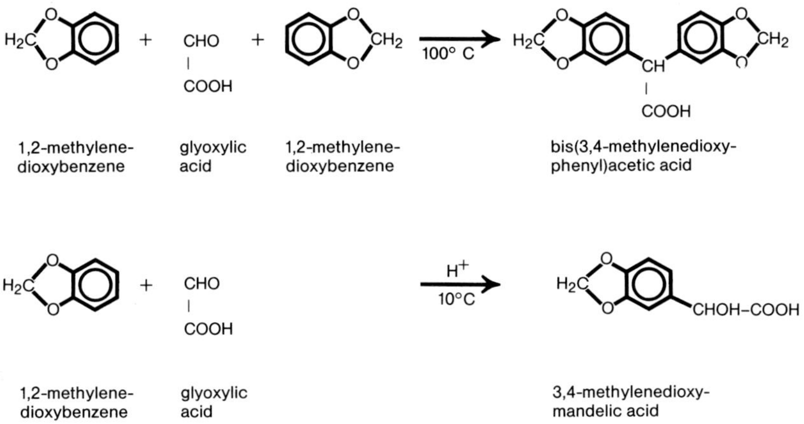

Fig. 5 Conversion of 1,2-methylenedioxybenzene and glyoxylic acid to the corresponding mono- and binuclear derivatives.

However, in order to synthesize 3,4-methylenedioxymandelic acid, the reaction has to be performed at much lower temperatures (Fig. 5). This automatically results in the reaction being no longer technologically viable due to considerable reduction in the reaction rate. This point is reviewed in more detail elsewhere (see symposium paper No. 5). Surprisingly, attempts to increase the reaction rate through the addition of a strong acid resulted in the binuclear compound no longer constituting the main product of the reaction. This finding clearly indicates that specific reaction conditions can be found for the phenol ether–aldehyde reaction such that the very reactive and labile mandelic acid intermediate (Fig. 5) can, in fact, be obtained in excellent yield. It was reported on an earlier occasion [6], that the water formed during the condensation reaction of phenols with aldehydes to form bis(hydroxyphenyl)alkanes should be removed as soon as it is formed. If this is not done, then the aldehydes are transformed into their corresponding aldols. This happens also in the reaction between phenol ethers and aldehydes (e.g., anisole and propionaldehyde). In this case too, highly specific reaction conditions were established under which aldol formation is hardly detectable and the required *para*-substitution is optimal.

Application of Intermediates to the Synthesis of Fragrance and Flavor Substances

The phenolic alcohols saligenin and 4-hydroxybenzyl alcohol are well-known as useful intermediates from which a large number of important fragrance and flavor substances are derived. Some examples of these conversions are shown in Figures 6 and 7. Considerable improvement in the synthesis of these compounds, through the use of a specific catalyst for the

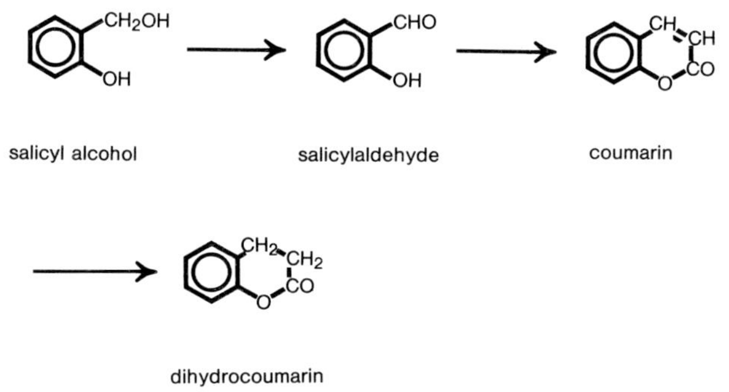

Fig. 6 Utilization of salicyl alcohol in the preparation of some fragrance and flavor substances.

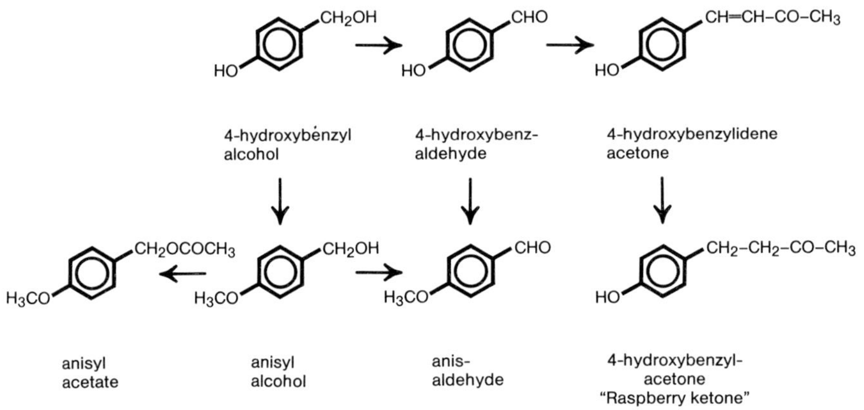

Fig. 7 Utilization of 4-hydroxybenzyl alcohol in the preparation of some fragrance and flavor substances.

oxidation of phenolic alcohols to their corresponding aldehydes, is described elsewhere (see paper No. 4). It is also well-known that the glycolic acids derived from glyoxylic acid and either phenols or monoethers of pyrocatechol are excellent starting materials for the synthesis of 4-hydroxybenzaldehydes [7]. A good illustration of this approach is the synthesis of vanillin via oxidation (using molecular oxygen under alkaline conditions) of the mandelic acid obtained from guaiacol and glyoxylic acid [8] (Fig. 8). The resulting substituted glyoxylic acid is then treated with mineral acid, which causes decarboxylation with the concomitant formation of vanillin. Acidic solutions of the mandelic acids can also be oxidized in the presence of metal salts capable of undergoing redox reactions [9] (Fig. 9). Very good yields of the resulting 4-hydroxybenzaldehydes can be obtained if these products are rapidly removed from the reaction mixture by using, for example, an extraction procedure.

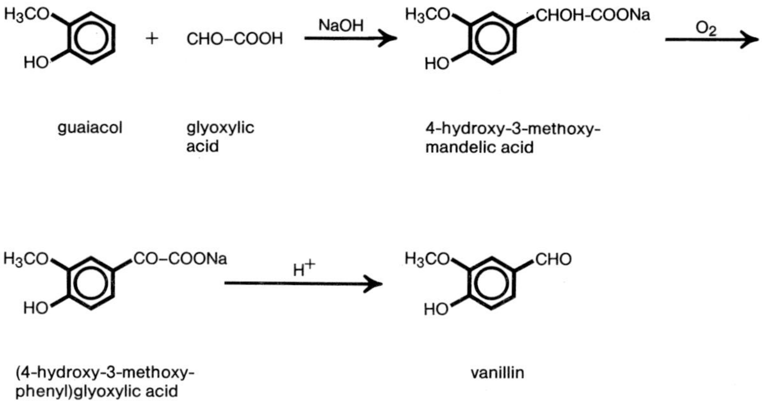

Fig. 8 Synthesis of vanillin via oxidation of the mandelic acid obtained from guaiacol and glyoxylic acid.

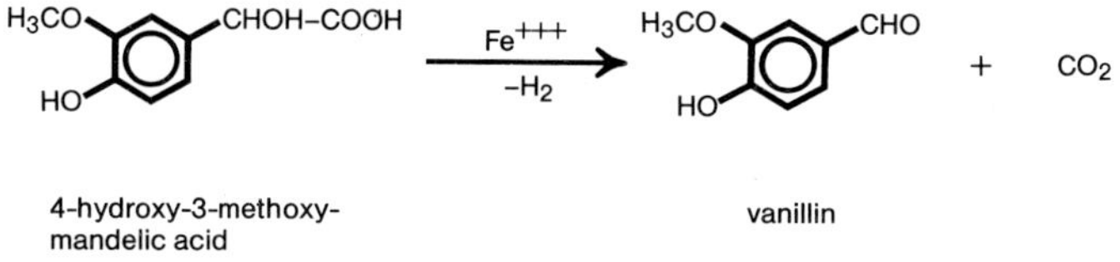

Fig. 9 An alternate means for the oxidation of 4-hydroxy-3-methoxymandelic acid to vanillin in the presence of metal salts.

Special conditions are required to oxidize (to their corresponding aldehydes) the mandelic acids obtained from phenol ethers. Figure 10 shows the specific and elegant transformation of 3,4-methylenedioxymandelic acid to heliotropin using nitric acid as the oxidizing agent (this procedure is described in greater detail in paper No. 5 of this symposium).

The bis(hydroxyphenyl)alkanes, which are obtained in good yields by condensation of phenols and aliphatic aldehydes (Fig. 2), have, until recently, received little attention as raw materials for the synthesis of fragrance and flavor substances. It has been reported in the literature that bis(hydroxyphenyl)alkanes undergo thermal cleavage under alkaline conditions, thereby yielding equimolar proportions of the alkenylphenol and phenol itself [10]. However, this procedure found practical application only in the cleavage of 2,2-bis(4-hydroxy-

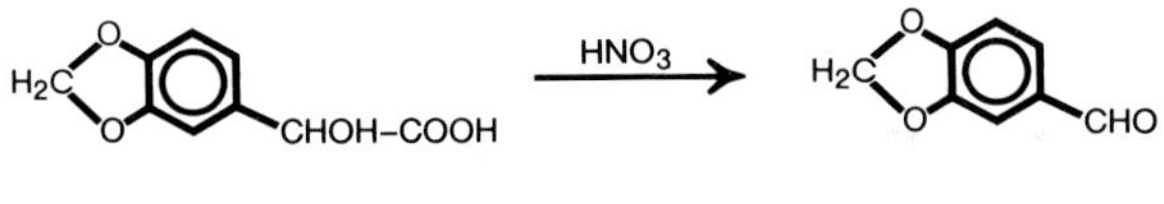

3,4-methylendioxy-
mandelic acid

heliotropin

Fig. 10 Oxidation of 3,4-methylenedioxymandelic acid to heliotropin with nitric acid. For further details of this procedure see paper no. 5 of this Symposium.

phenyl)propane (obtained from phenol and acetone) to yield 4-isopropenylphenol and phenol. By further developing this procedure, originally described by Krimm and Schnell [2], a successful synthesis of isoeugenol was developed. This synthesis [11], using the previously mentioned bisphenol prepared from guaiacol and propionaldehyde, is illustrated in Figure 11.

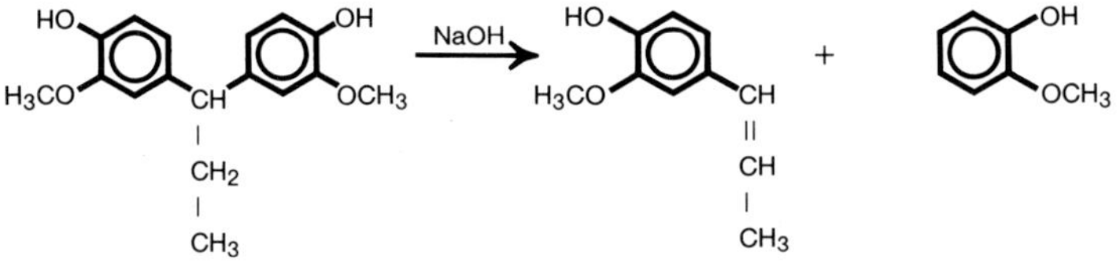

1,1-bis(4hydroxy-3-methoxy-
phenyl)propane

isoeugenol

guaiacol

Fig. 11 Cleavage of 1,1-bis(4-hydroxy-3-methoxyphenyl)propane to isoeugenol and guaiacol.

Attempts to obtain 4-hydroxypropenylbenzene by similar alkaline cleavage of the bisphenol obtained from phenol and propionaldehyde were also successful, as shown in Figure 12 [12]. Attempts to convert 4-hydroxypropenylbenzene to the methyl ether (anethole) by using only methanol were unsuccessful, since the alkenylphenol was inclined to polymerize under the acidic reaction conditions necessary. It therefore represented an important step forward when it was discovered that 1,1-bis(4-methoxyphenyl)propane can be cleaved under very specific acidic reaction conditions to yield equimolar quantities of anisole and anethole. This procedure is shown schematically in Figure 13, and is described in detail elsewhere [13], (see also paper No. 7 of the symposium).

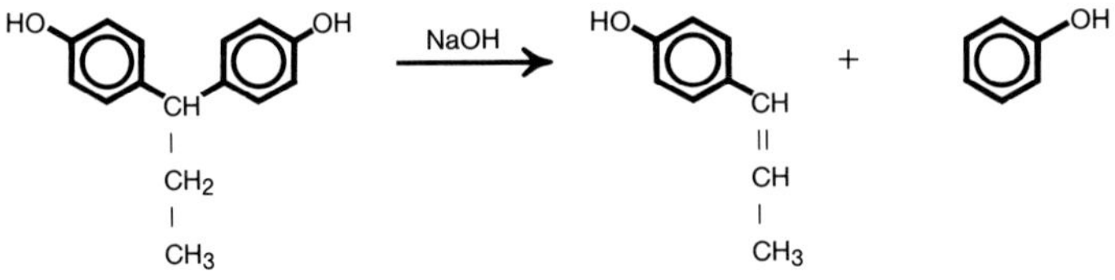

1,1-bis(4-hydroxy-
phenyl)propane

4-hydroxypropenyl-
benzene

phenol

Fig. 12 Cleavage of 1,1-bis(4-hydroxyphenyl)propane to 4-hydroxypropenylbenzene and phenol.

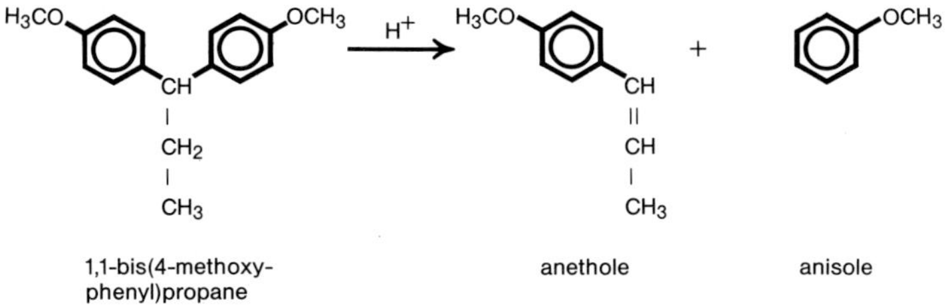

Fig. 13 Cleavage of 1,1-bis(4-methoxyphenyl)propane to anethole and anisole.

This short review, intended to summarize but one research area in which Haarmann & Reimer has been active, also illustrates how the detailed study of a "classical" organic reaction can lead to development of new economical processes for the production of some very important fragrance and flavor substances, which are either phenols themselves or derivatives thereof.

Acknowledgement

I would sincerely like to thank my co-workers, especially Dr. Steuer, Dr. Mölleken and Alfred Krempel, for their excellent collaboration and team work, without which the work described here would not have been possible.

References

[1] Kirk-Othmer, Encyclopedia of Chemical Technology, 2nd Edition Vol. 15, pp. 176–208, Wiley-Interscience, New York, London, Sydney, Toronto, 1967.
[2] Schnell, H., and H. Krimm, Angew. Chemie **75**, 662–668 (1963).
[3] Esso Research and Engineering Co., Brit. Patent 842209 (July 20, 1960).
[4] Wegler, R., and H. Herlinger in Houben-Weyl, Methoden der Organischen Chemie, 4. Auflage, Vol. 14/2, pp. 193–291, Georg Thieme Verlag, Stuttgart, 1963.
[5] Bauer, K., R. Mölleken, K. Wedemeyer, H. Fiege (Haarmann & Reimer GmbH), German Patent Application P 2 915 216.7.
[6] Bauer, K., R. Mölleken and R. Hopp, VII. International Congress of Essential Oils, Kyoto, 1977, pp. 264–266.
[7] Bayer, O., in Houben-Weyl, Methoden der Organischen Chemie, 4. Auflage, Vol. 7/1, pp. 312–313, Georg Thieme Verlag, Stuttgart, 1954.
[8] Nagai, S., T. Enomiya, T. Nakamura (Ube Industries, Ltd.), Japan Kokai 76 128 934.
[9] Bauer, K., W. Steuer (Haarmann & Reimer GmbH), German Patent 2 115 551 (Oct. 12, 1972).
[10] Krimm, H., H. Schnell (Farbenfabriken Bayer AG), German Patent 1 235 894 (March 9, 1967).
[11] Bauer, K., R. Mölleken, R. Hopp (Haarmann & Reimer GmbH), German Patent 2 418 973 (Sept. 25, 1975).
[12] Unpublished.
[13] Bauer, K., R. Mölleken (Haarmann & Reimer GmbH), German Patent 2 418 974 (Sept. 18, 1975).

4. FURTHER DEVELOPMENT OF A CLASSICAL PROCESS FOR THE SYNTHESIS OF AROMATIC HYDROXYALDEHYDES

H. Fiege*, K. Wedemeyer*, K. A. Bauer**, A. Krempel** and R. G. Mölleken**

Summary A very selective process for the synthesis of salicylaldehyde and *p*-hydroxybenzaldehyde was developed in which phenol, formaldehyde and oxygen are the starting materials. The process is not only more economical, but also operates at a much lower pollution level than all the other known industrial processes for the production of salicylaldehyde and *p*-hydroxybenzaldehyde. Furthermore, it is possible to regulate the relative proportions of the two isomeric aldehydes within certain limits according to market requirements. The process may also be employed for the production of a large number of substituted aromatic hydroxyaldehydes which could be of great interest in the fragrance and flavor field or in other areas.

Introduction

A large number of syntheses have been developed to provide aromatic hydroxyaldehydes. It may therefore come as a surprise to find that there is still a genuine interest in looking for improved processes in this field. However, the increasing necessities of environmental protection call for new ideas which, while making due allowance for ecological considerations, will nevertheless ensure economical production of key aromatic hydroxyaldehydes on a long-term basis.

Most important among the aromatic hydroxyaldehydes, in terms of quantity, is salicylaldehyde, and one of the major uses of this compound lies in the manufacture of aroma chemicals such as coumarin [1]. Other uses of salicylaldehyde are in the synthesis of fuel and oil additives, pesticides and dyestuffs. Additionally, there are numerous minor applications.

Syntheses of Salicylaldehyde

There are over 30 processes for the synthesis of salicylaldehyde. At present three of these are used commercially: the Reimer-Tiemann process; the amalgam reduction of salicylic acid; and the phenyl metaborate/saligenin process.

---

* H. Fiege and K. Wedemeyer, Bayer AG, 5090 Leverkusen, West Germany.
** K. A. Bauer, A. Krempel and R. G. Mölleken, Haarmann & Reimer GmbH, 3450 Holzminden, West Germany.

The Reimer-Tiemann Process

The first of the three well established processes for the production of salicylaldehyde dates back to K. Reimer, a cofounder of Haarmann & Reimer, who published this synthesis together with F. Tiemann in 1876 [2] (Fig. 1). This process is still commercially attractive for several reasons. Firstly, it consists essentially of only one chemical step (excluding acidification of the reaction mixture). Secondly, the raw materials, phenol, chloroform, concentrated caustic soda solution and sulfuric or hydrochloric acid, are readily accessible. Finally, the process permits the co-production of *p*-hydroxybenzaldehyde, which is useful for the manufacture of such aroma chemicals as raspberry ketone, anisaldehyde and anisic alcohol, as well as pesticides and drugs. It is not possible to produce *p*-hydroxybenzaldehyde by the two other processes to be described below.

REIMER-TIEMANN process

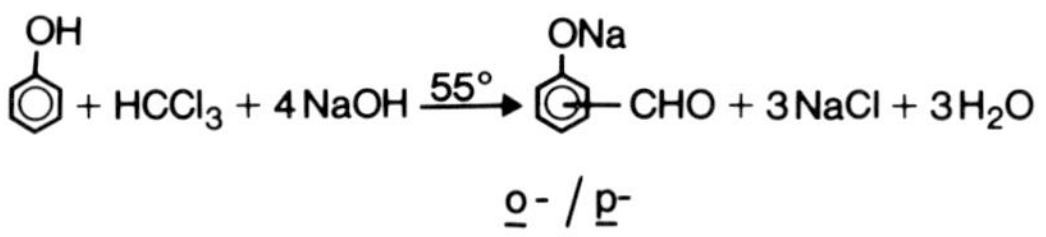

	with	without solvent
yield, rel. phenol	~ 80	⩽ 60
yield, rel. $CHCl_3$	~ 50	
o-/p- ratio	65 : 35	85 : 15
kg tar/kg aldehyde	~ 0,2	⩾ 0,5
kg salts/kg aldehyde	⩾ 6	

Fig. 1 The Reimer-Tiemann process for the production of salicylaldehyde and *p*-hydroxybenzaldehyde.

Among the drawbacks of the Reimer-Tiemann process are the yields. Under the best conditions, where phenol is used in great molar excess relative to chloroform, and where the reaction is conducted in the presence of a solvent such as methanol, the yield of salicylaldehyde plus *p*-hydroxybenzaldehyde may be on the order of 80% based on reacted phenol, but only about 50% relative to chloroform [3]. The salicylaldehyde: *p*-hydroxybenzaldehyde proportions are approximately 65:35 in the presence of solvent, but only about 85:15 without solvent, and in this case the yield is roughly 60%. Furthermore, without solvent far more tar-like products are obtained, and this complicates post-reaction processing, especially the isolation of pure *p*-hydroxybenzaldehyde. A prerequisite of the process is, of course, the recovery of the excess phenol and any solvent from the aqueous mixture. These operations involve substantial capital and energy expenditure.

The effluent resulting from the process also poses an ecological problem. Beside organic pollutants, the effluent contains a high salt content (i.e., more than 6 kg of a sodium chloride/sodium sulfate mixture are obtained per kg aldehyde when acidifying with sulfuric acid). This large quantity of salt is, to some extent, due to the excess of caustic soda solution used.

Thus, while the Reimer-Tiemann process is chemically sound, it involves substantial capital and energy expenditure, and it requires large amounts of chemicals which encumber its technology and affect the environment.

The Amalgam Reduction of Salicylic Acid

An alternate procedure for the preparation of salicylaldehyde involves the sodium amalgam reduction of sodium salicylate in the presence of boric acid and sodium bisulfite, with the sodium amalgam being produced electrochemically either separately or *in situ* at a mercury cathode [4, 5] (Fig. 2). The boric acid is required because it is not the salicylic acid itself which is reduced directly by the sodium, but rather the intermediate boric acid-salicylic acid ester-anhydride (shown in brackets in Fig. 2). This explains why *p*-hydroxybenzaldehyde is not accessible from the corresponding *p*-hydroxybenzoic acid by this process. To maintain the optimal pH during the reaction sulfuric acid is added, and the resultant salicylaldehyde product is converted to the corresponding bisulfite addition compound to prevent further reduction to the alcohol. The bisulfite adduct is then decomposed by steam heating the reaction mixture, and the salicylaldehyde is recovered from the steam distillate. The yield of the process is about 60%.

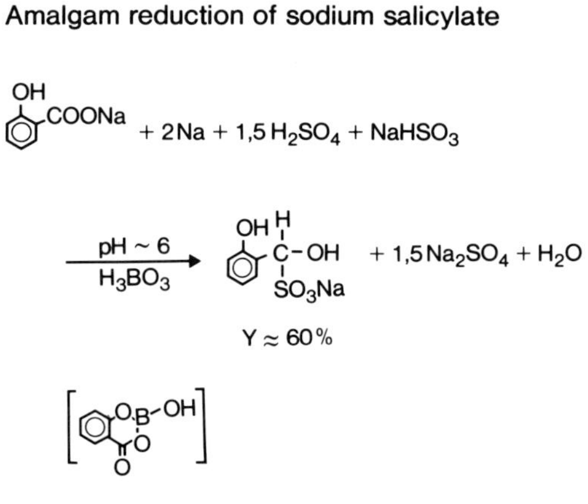

Fig. 2 The amalgam reduction of sodium salicylate to salicylaldehyde. The intermediate boric acid-salicylic acid ester-anhydride is shown in brackets.

After the steam distillation step, there remains an aqueous sodium sulfate solution from which the non-reacted, relatively costly salicylic acid, as well as the boric acid and the sodium bisulfite, cannot be recovered without losses (even at high technological expenditure). This means that this elegant process is ecologically unsatisfactory, and, furthermore, involves great cost and technological expenditure because of the recovery operations.

The Phenylmetaborate/Saligenin Process

A third commercially adopted approach to the synthesis of salicylaldehyde is illustrated in Figure 3. This process [6] is based upon the reaction of phenol with boric acid to give

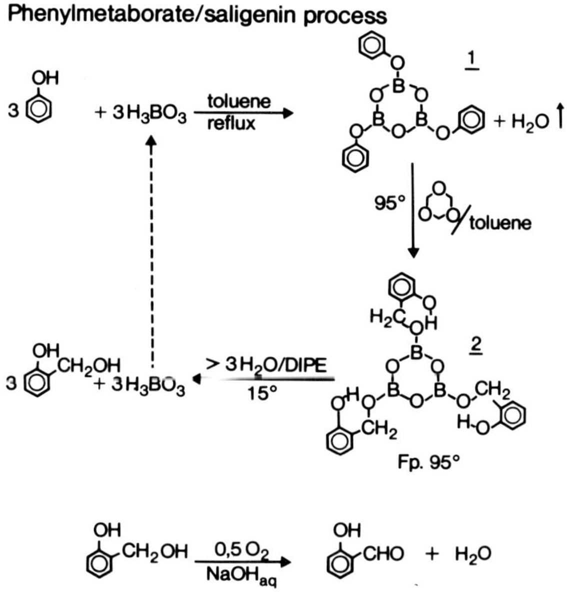

Fig. 3 The phenylmetaborate/saligenin process for the synthesis of salicylaldehyde.

triphenylmetaborate (*1*). The water released during this esterification is removed by azeotropic distillation with toluene. The triphenylmetaborate (*1*) is then hydroxymethylated with trioxymethylene in toluene in the absence of water at 95 °C, and without the isolation of any intermediates. In this operation the boric acid has two functions; it directs the hydroxymethyl groups into the *ortho*-position relative to the hydroxyl group of the phenol, and it stabilizes the resulting 2-hydroxymethylphenol (the so-called saligenin) in the form of saligeninmetaborate (*2*). The result is a saligeninmetaborate/toluene crystal slurry which is hydrolyzed with excess water. The addition of diisopropylether (DIPE) precipitates the boric acid which can be filtered off and, after washing with toluene, recycled [7]. The saligenin in the toluene/diisopropylether phase partially crystallizes after the diisopropylether has been distilled off under mild conditions. The yield of the resultant saligenin is on the order of 60 % to 70 % relative to phenol or trioxymethylene. After washing with toluene and removal of solvent residue, the saligenin is dissolved in caustic soda solution and oxidized with oxygen in the presence of noble metal catalysts to give salicylaldehyde [8, 9].

It can easily be seen that the phenylmetaborate/saligenin process is restricted to the production of salicylaldehyde, and is quite costly because of the many chemical steps involved and the necessary handling of crystalline masses. In addition, some boric acid is bound to be lost in the effluent.

Development of a Classical Procedure

The three processes described above, as well as a number of other possible approaches to the synthesis of salicylaldehyde and/or *p*-hydroxybenzaldehyde, were subjected to critical comparison and experimental verification. The results showed that the best approach would be a base catalyzed direct reaction of phenol with formaldehyde to give 2- and 4-hydroxymethylphenol, followed by noble metal-catalyzed oxidation of the hydroxymethylphenols using oxygen to yield the respective phenolaldehydes (Fig. 4). The attractiveness of this concept was based on the inexpensive readily accessible raw materials, the small number of chemical steps involved, the possibility of producing *p*-hydroxybenzaldehyde at the same time, and the chance to arrive at an ecologically satisfactory process.

Historically, the two reaction steps proposed (Fig. 4), have, in principle, been known for quite a long time. A patent application for the hydroxymethylation, which became known under the name of the Lederer-Manasse reaction [10], had been filed by Farbenfabriken Bayer as early as 1894 [11], and the oxidation of saligenin to salicylaldehyde with oxygen using platinum catalysts had been reported by Piria even earlier [12]. However, no data on yields were given in either case.

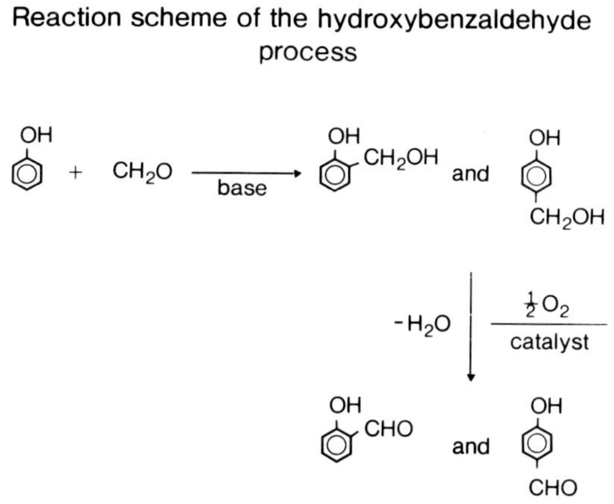

Fig. 4 Reaction scheme for the hydroxybenzaldehyde process based on the direct reaction of phenol with formaldehyde.

Initially, we were somewhat sceptical about this concept because of the chemical and thermal instability of the hydroxymethylphenols, which are well-known as extremely reactive intermediate stages in the production of phenol formaldehyde resins [13]. Thus, we were led to expect numerous problems with the production, isolation, oxidation and analysis of these compounds. However, after weighing all the pros and cons, we decided to launch a joint study of the various problems with the tasks being divided among the Bayer and the Haarmann & Reimer groups.

While the formaldehyde/phenol reaction (Fig. 5) appears to be simple, it actually comprises a complex system of reactions starting with the depolymerization of paraformaldehyde to give the formaldehyde monomer, which in turn reacts with phenol to form 2- and 4-hydroxymethylphenol (2- and 4-HMP). Both methylol compounds are capable of reacting further with formaldehyde, via the intermediate 2,6- and 2,4-dihydroxymethylphenol (2,6- and 2,4-DHMP), to provide 2,4,6-trihydroxymethylphenol (2,4,6-THMP) (Fig. 5). Additionally, all

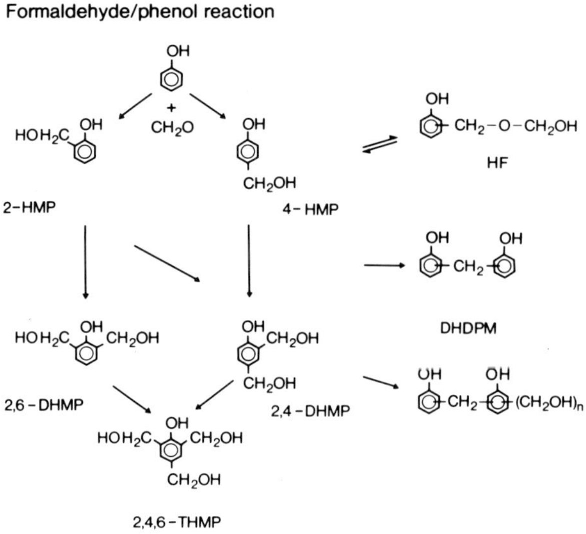

Fig. 5 Products of the formaldehyde/phenol reaction. 2- and 4-HMP represent 2- and 4-hydroxymethylphenol. 2,6- and 2,4-DHMP represent 2,6- and 2,4-dihydroxymethylphenol. 2,4,6-THMP is equivalent to 2,4,6-trihydroxymethylphenol. HF equals hemiformal, while DHDPM represents dihydroxydiphenylmethane.

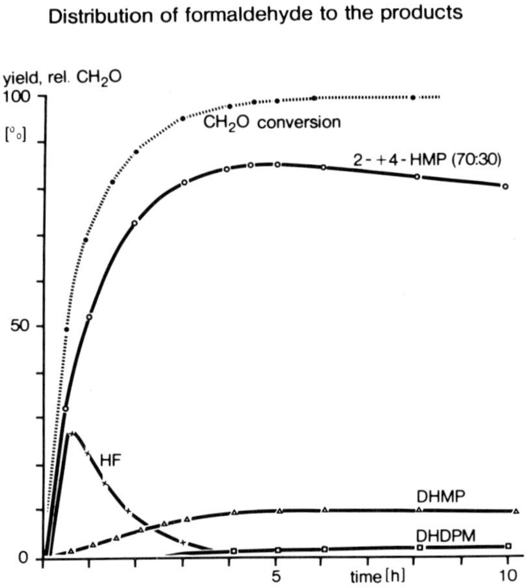

Fig. 6 Distribution of formaldehyde into the products of the formaldehyde/phenol reaction. For simplicity, the structurally isomeric compounds are grouped together in each curve. Abbreviations are the same as in Fig. 5.

hydroxymethylphenols are capable of reversibly forming hemiformals with additional formaldehyde [14] and condensing further, either with excess phenol to form dihydroxydiphenylmethanes, or with each other to form polynuclear hydroxymethylphenols (Fig. 5). A wide variety of other reactions are also possible.

According to the reaction scheme an excess of phenol promotes the formation of 2- and 4-hydroxymethylphenol, and in practice, phenol : formaldehyde molar proportions greater than 5 : 1 have proved useful. Figure 6 shows the percent distribution of formaldehyde between the individual products as a function of time with a large excess of phenol at 60 °C, and using potassium hydroxide as a catalyst. Considerable formation of hemiformal takes place at the beginning of the reaction. However, it drops below the detectable limit when the yield of 2- and 4-hydroxymethylphenol has reached its maximum at about 85 %. At this time there still is a slight increase in the quantity of undesired dihydroxymethylphenols. The total formaldehyde conversion is over 98 % in the region where the HMP yield is greatest (Fig. 6). With increasing time, the yield of the two desired HMP main products decreases, a result of condensation to polynuclear phenols.

At higher temperatures, the phenol/formaldehyde reaction to yield 2- and 4-hydroxymethylphenols can be conducted at much faster rates than that shown in Figure 6; however, it is critical that the reaction be terminated at the proper time. The termination time of the reaction is, of course, not only a function of temperature, but also depends on such parameters as the phenol : formaldehyde ratio, the type and concentration of catalyst, the type of formaldehyde, and the water content of the system. We determined the quantitative relationships between these parameters during the course of extensive process oriented kinetic studies, which also included the decomposition rates during post-reaction processing. With knowledge of the relationships, we are able to supervise this complex phenol/formaldehyde reaction analytically using a relatively simple physical technique, and thereby conduct the process under optimum conditions. Thus, the 2- and 4-hydroxymethylphenol yields obtainable under economical continuous operation amount to more than 85 % of theoretical.

The proportions of 2- and 4-hydroxymethylphenol, which are important for the flexibility of the process, are primarily determined by the type of catalyst used, and the ratio of 2-HMP to 4-HMP can be adjusted between 99 : 1 and approximately 50 : 50 (Table 1). Ratios between the extremes can be obtained with catalyst mixtures. The higher the charge and the smaller the radius of the catalyst ion, the higher is the content of 2-hydroxymethylphenol. On the other hand, very

Tab. 1 **Effect of catalyst on the 2-HMP : 4-HMP ratio**

Catalyst	2-HMP : 4-HMP
$Mg(OAc)_2$	99 : 1
$Zn(OAc)_2$	98 : 2
$Ca(OH)_2$	95 : 5
LiOH	90 : 10
NaOH	80 : 20
KOH	70 : 30
KOH + polyglycol ether	50 : 50

small quantities of certain polyglycol ethers cauce the 4-hydroxymethylphenol content to increase substantially (Table 1). The necessary catalyst concentrations are as low as 10^{-3} moles per mole phenol, which means that they are substantially below the recommended levels found in the literature [14, 15] and insignificant in cost. With regard to the mechanism of hydroxymethylation, it is assumed from the literature that a chelate mechanism plays a role in *ortho*-substitution, whereas *para*-substitution takes an exclusively ionic course [16].

The isolation of 2- and 4-hydroxymethylphenol from the reaction mixtures obtained in the presence of basic catalysts is difficult, and most methods attempted were either not viable on a commercial scale or involved unacceptable losses of yield. However, after extensive studies we arrived at a technologically workable, economical approach to this problem in the form of a special multi-stage countercurrent extraction between water and solvents such as toluene and diisopropylether [17]. This process allows the isolation of both 2- and 4-hydroxymethylphenol in nearly quantitative yield, and practically free of by-products. The products are obtained in the form of an aqueous-alkaline solution which can be oxidized directly to give the relevant phenolaldehydes. The extraction procedure also allows for the quantitative recovery of the excess phenol which is recycled, and it provides the by-products with properties that allow their marketing as novolaks. The post-reaction processing takes place under very mild conditions and requires no handling of the thermally unstable hydroxymethylphenols in free form.

According to the literature, it is possible to oxidize aqueous-alkaline 2-hydroxymethylphenol with oxygen at room temperature and in the presence of palladium [8], or preferably platinum on carbon contacts [9, 18], to form salicylaldehyde. When we started our work, however, the state of technology was such that oxidation times of several hours were necessary, in spite of large contact quantities and precious metal concentrations of 5–10 %. Even under the best technical conditions the quantity of platinum metal used was greater than 500 mg per mole hydroxymethylphenol, and the salicylaldehyde yield was only about 80 %. In addition, approximately 17 % tar and, because of over-oxidation, 2–3 % salicylic acid were obtained. The reuse of the contacts was greatly limited by the heavy tar formation. A process on this basis appeared to be completely impractical because of the high platinum costs.

We succeeded in drastically improving the activity and selectivity of the platinum-carbon contacts by introducing certain non-noble metals, such as very small quantities of lead [19]. This makes it possible, for example, to oxidize 2-hydroxymethylphenol with oxygen in an aqueous alkaline solution at normal pressure on a contact which contains only 1 % platinum, and in as short a time as 20 minutes. In this case, the salicylaldehyde yield is over 98 %, and there is no appreciable formation of tar or salicylic acid. As shown by the oxidation curve for the conversion of 2-HMP to salicylaldehyde (Fig. 7), oxidation stops after the stoichiometric quantity of oxygen has been taken up. The oxidation of 4-HMP to *p*-hydroxybenzaldehyde behaves in the same way. Without the introduction of lead there is practically no oxygen up-take under these conditions. The hydroxymethyl compounds, as well as the phenolaldehydes formed, are completely dissolved as phenolates in the aqueous alkaline medium, and therefore no problems are encountered during the oxidation with oxygen and the ensuing separation of the catalyst. The catalyst can be reused so often that the consumption of platinum metal is below 1 mg per mole hydroxymethylphenol, and this does not even take into account the credit for the platinum contained in the used contact.

After separation of the catalyst, the aldehydes are released by acidification with sulfuric acid and the salicylaldehyde is stripped from the aldehyde mixture with steam. The *p*-hydroxybenz-

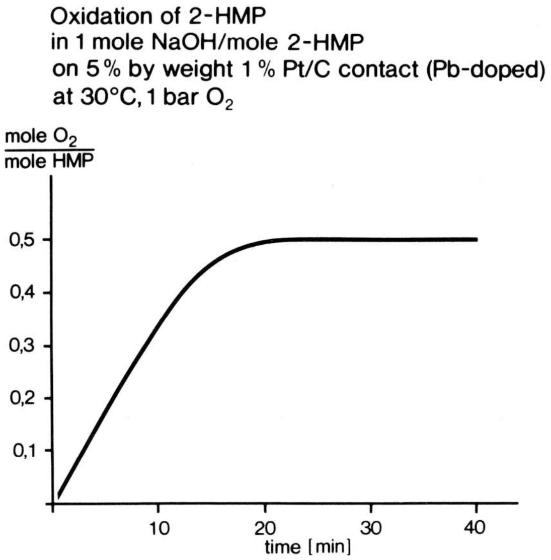

Fig. 7 Oxidation curve for the conversion of 2-hydroxymethylphenol to salicylaldehyde. The oxidation of 2-HMP was carried out in 1 mole NaOH per mole of 2-HMP on 5% by weight 1% Pt/C contact (Pb-doped) at 30 °C, 1 bar O_2.

aldehyde crystallizes out in the bottom of the steam distillation column and is filtered off. After drying, it easily conforms with commercial quality. The effluent, which essentially contains only sodium sulfate, is further purified in a biological treatment plant. Where no such facility exists, the organic matter can also be removed by extraction before it is discharged into the sewage system.

We believe that the process described here has several advantages over other commercially useful salicylaldehyde syntheses. It is much more environmentally compatible than any other method. It allows flexible control of the salicylaldehyde/*p*-hydroxybenzaldehyde proportions geared to market demands, and it supplies both aldehydes at production costs lower than those of other processes. Bench scale trials have proved that this procedure is generally viable, and a pilot scale trial is well advanced.

Further Applications of the Oxidation Catalyst

Further studies showed that the use of the oxidation contacts described here is not limited to the synthesis of salicylaldehyde and *p*-hydroxybenzaldehyde but rather allows highly selective production of a wide variety of aromatic hydroxyaldehydes [19]. Figure 8 illustrates some examples of alkyl substituted aromatic hydroxyaldehydes prepared by oxidation of the correspondingly substituted hydroxymethyl compounds. The yields were, in most cases, quantitative. In cases where an oxidation was not possible because of the insolubility of the phenol in caustic soda solution, the problem could be overcome by the addition of a suitable

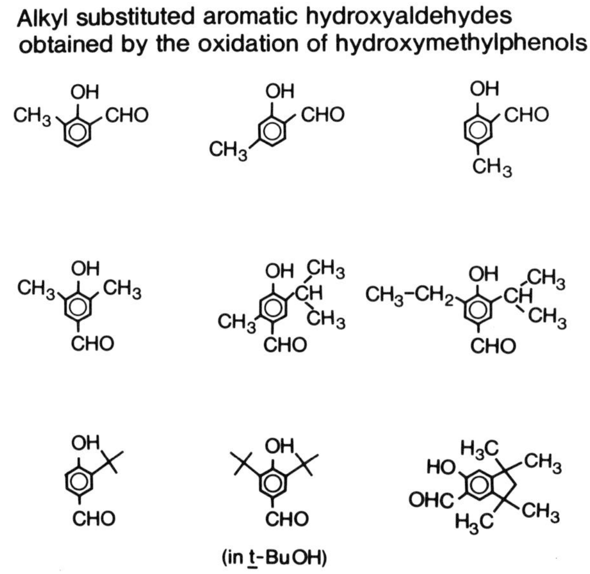

Fig. 8 Alkyl substituted aromatic hydroxyaldehydes obtained by the oxidation of the corresponding hydroxymethylphenols.

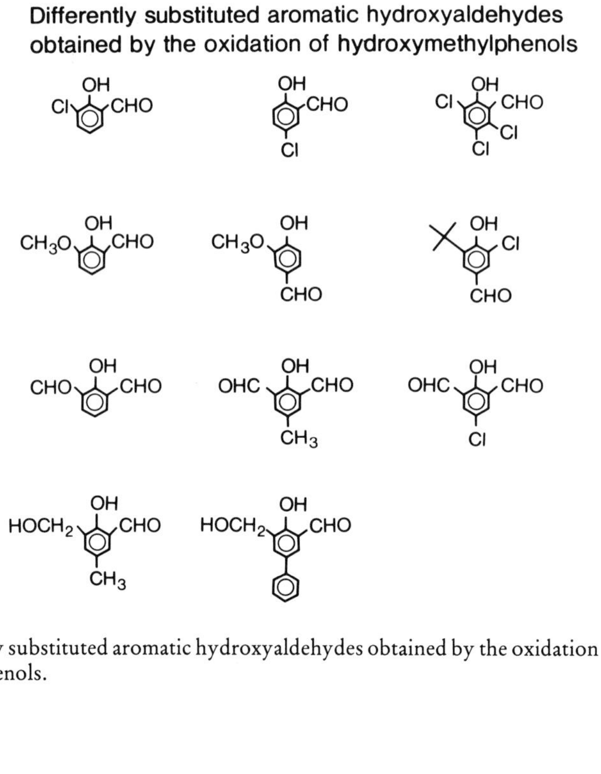

Fig. 9 Differently substituted aromatic hydroxyaldehydes obtained by the oxidation of the corresponding hydroxymethylphenols.

solvent. For example, the oxidation of 2,6-di-*t*-butyl-4-hydroxymethylphenol to 3,5-di-*t*-butyl-4-hydroxybenzaldehyde (Fig. 8) was not possible in aqueous caustic soda solution, but could be carried out in good yield in the presence of some *t*-butanol. The selection of aryl, halogen, and alkoxy substituted phenolaldehydes shown in Figure 9 were also prepared from the corresponding hydroxymethyl compounds, illustrating the broad range of application of the process. Furthermore, dihydroxymethylphenols can be oxidized to the corresponding dialdehydes, and in some cases also to the hydroxymethylphenolaldehydes, by this procedure.

Since it is often considerably easier to produce the substituted hydroxymethylphenols from the corresponding substituted phenols than it is to produce 2- and 4-hydroxymethylphenol from phenol, we consider our procedure to be a versatile approach to the synthesis of aldehydes which otherwise are not readily accessible, and which therefore have found little application.

References

[1] Voigt, H., in Ullmanns Encyclopädie der technischen Chemie, Vol. IX, p. 637–640, 4th. Ed., Verlag Chemie, Weinheim 1975.
[2] Reimer, K., Ber. 9, 423 (1876); Reimer, K., and F. Tiemann, Ber. **9**, 824, 1268 (1876).
[3] U.S. Pat. 3 365 500 (1964) to Dow Chemical Co.
[4] Weil, H., Ber. **41**, 4147–8 (1908).
[5] Udupa, K. S., G. S. Subramanian and H. V. K. Udupa, The Industrial Chemist **39**, 238–241 (1963).
[6] U.S. Pat. 3 290 352, 3 290 393 (1963) to Rhône-Poulenc S. A.
[7] U.S. Pat. 3 868 424 (1970) to Rhône-Poulenc S. A.
[8] U.S. Pat. 3321526 (1963) to Rhône-Poulenc S. A.
[9] U.S. Pat. 4026950 (1976) to Rhône-Poulenc Industries.
[10] Walker, J. F., Formaldehyde, p. 331, 3rd. Ed., Reinhold Publishing Co., New York 1964.
[11] Ger. Pat. 85588 (1884) to Farbenfabriken Bayer.
[12] Piria, R., Liebigs Ann. Chem. **56**, 42 (1845–49).
[13] Wagner, H., and H. F. Sarx, Lackkunstharze, p. 33–37, C. Hanser Verlag, München 1971.
[14] Zavitsas, A. A., R. D. Beaulieu and J. R. Le Blanc, J. Polym. Sci., (A-1) **6**, 2541–59 (1968).
[15] Partansky, A. M., Am. Chem. Soc., Div. Org. Coatings Plast. Chem. **27**, 115–24 (1967).
[16] Becker, H. G. O., Einführung in die Elektronentheorie organisch-chemischer Reaktionen, p. 307, 3rd. Ed., VEB Deutscher Verlag der Wissenschaften, Berlin 1974.
[17] E.U. Pat. 0 000 165 (1978) to Haarmann & Reimer GmbH.
[18] U.S. Pat. 3 673 257 (1970) to Tenneco Chemicals, Inc.
[19] U.S. Pat. 4 119 671 (1977) to Bayer AG.

5. A NEW METHOD FOR THE PRODUCTION OF HELIOTROPIN

R. G. Mölleken* and K. A. Bauer*

Summary The common starting material for the industrial synthesis of heliotropin is the natural product safrole, and, since safrole is still reasonable in price, there has been little demand for completely synthetic heliotropin. However, the recent advent of a large scale industrial process for the production of pyrocatechol, which is readily converted to 1,2-methylenedioxybenzene (*1*), has greatly changed this situation, and the total synthesis of heliotropin (*4*) is now economically viable.

Scheme 1

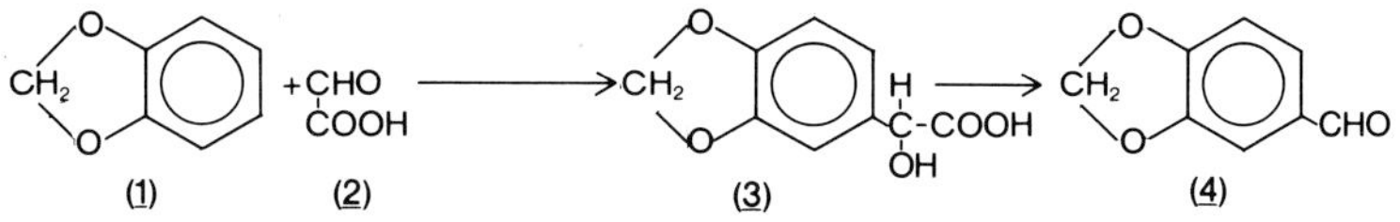

This new synthesis involves the condensation of 1,2-methylenedioxybenzene (*1*) and glyoxylic acid (*2*) to 3,4-methylenedioxymandelic acid (*3*), which can be oxidatively decarboxylated in good yield to heliotropin (*4*) by treatment with dilute nitric acid.

Introduction

The importance and the long history of heliotropin are reflected by the number of names for this interesting compound. Seven synonyms for heliotropin have been reported in the literature, the newest being 1,3-benzodioxole-5-carboxaldehyde [1] (Fig. 1). Heliotropin was discovered by Fittig and Mielch [2] while studying the hydrolysis and oxidation reactions of piperine, the main alkaloid of black pepper *(Piper nigrum)* (Fig. 2). Heliotropin itself occurs in small amounts in the essential oils of many tropical and subtropical plants [3]. However, it is never the main constituent of these oils, which explains why industrial production of heliotropin from essential oils is unknown.

* R. G. Mölleken and K. A. Bauer, Haarmann & Reimer GmbH, 3450 Holzminden, West Germany.

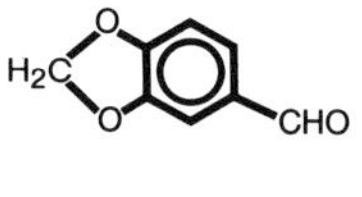

heliotropin

Piperonal
Piperonyl aldehyde
Protocatechinaldehyde methylene ether
Dioxymethylene protocatechine aldehyde
3,4-Methylenedioxybenzaldehyde
1-Aldehydo-3,4-methylenedioxy benzene
1,3-Benzodioxole-5-carboxaldehyde

Fig. 1 Synonyms for heliotropin.

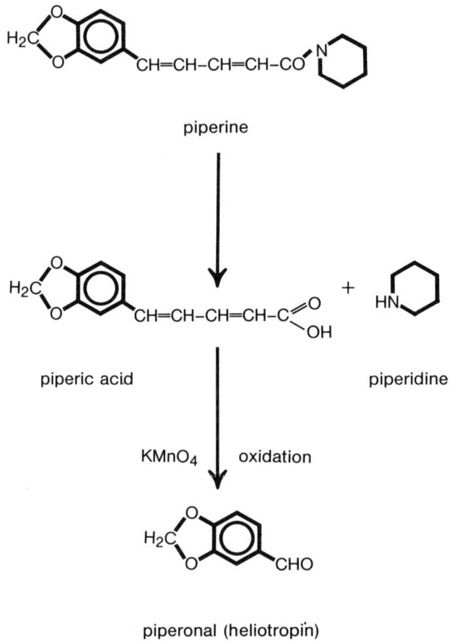

Fig. 2 Studies on the hydrolysis and oxidation of piperine led to the discovery of heliotropin by Fittig and Mielch [2].

Early Syntheses of Heliotropin

Eykman [4] and Ciamician and Silber [5] were the first to achieve a practical synthesis of heliotropin, by the oxidation of isosafrol (Fig. 3). Various oxidizing agents have since been used in the manufacture of heliotropin, most with good results (Fig. 3). Thus, for a long time isosafrol has been the preferred starting material for the production of heliotropin, and this procedure continues to be of considerable industrial importance. Safrol is still a relatively economical raw material for the production of isosafrol and is, at the moment, still available in sufficient quantities from camphor oil and Chinese Sassafras oils [6]. Nevertheless, attempts have been made to produce completely synthetic heliotropin, generally from pyrocatechol which is the most popular starting material.

Thus, the Russian researchers Lasskina *et. al.* [7] and Jelissejewa *et. al.* [8] have successfully synthesized heliotropin via the chloromethylation and/or hydroxymethylation of 1,2-

methylenedioxybenzene (Fig. 4). The chloromethyl and hydroxymethyl intermediates were not isolated in this process and the overall yield amounted to some 25 % relative to pyrocatechol. Bidlo *et. al.* [9] isolated the chloromethylated intermediate which was then subjected to the Sommelet reaction in the presence of hexamethylenetetramine, thereby yielding heliotropin (Fig. 4).

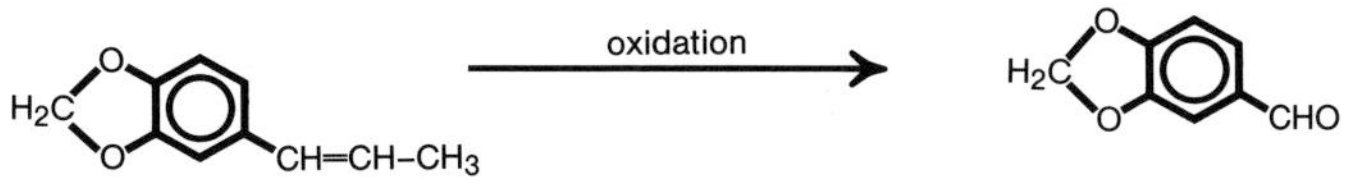

isosafrol heliotropin

oxidising agent	yield %	literature
HNO_3	traces	4
$KMnO_4$	80	4
H_2O_2/V_2O_5	68	5
O_2/γ-rays	high	6
CrO_3/H_3PO_4	87	7
$Na_2Cr_2O_7/Mn^{IV}$	high	8
O_3	96	9

Fig. 3 Synthesis of heliotropin by the oxidation of isosafrol.

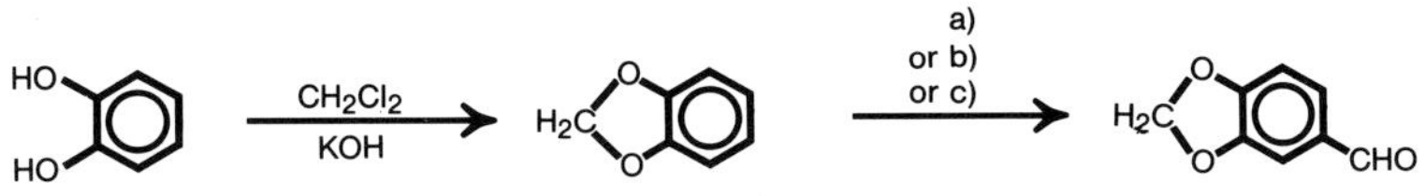

pyrocatechol 1,2-methylene-dioxybenzene heliotropin

a) $HCHO/HCl$, $m\text{-}O_2N\text{-}C_6H_4\text{-}SO_3H$
b) hexamethylenetetramine/4-nitrosodimethylaniline
c) $HCHO/HCl$, hexamethylenetetramine (Sommelet)

Fig. 4 Syntheses of heliotropin from pyrocatechol.

A procedure patented by Ube Industries [10] provided a considerable improvement upon these older processes (Fig. 5). Under alkaline conditions, glyoxylic acid undergoes an addition reaction with pyrocatechol to yield 3,4-dihydroxymandelic acid, which is oxidized to the corresponding keto carboxylic acid. The keto carboxylic acid, upon acidification, decarboxylates to yield protocatechuic aldehyde which is converted to heliotropin. The efficient conversion of protocatechuic aldehyde to heliotropin can be achieved by a new procedure belonging to Brichima SpA [11], such that the overall process yield amounts to some 50 %.

A Japanese group [12] applied the classical Vilsmeier aldehyde synthesis of heliotropin, by a procedure involving the reaction of 1,2-methylenedioxybenzene with N-methyl formanilide using phosphorous oxychloride or phosgene as condensing agent (Fig. 6). The by-product N-methyl aniline can easily be recovered and treated with formic acid to regenerate N-methyl formanilide, which is recycled. The overall yield is supposed to amount to some 77 %, but this could not be confirmed in our laboratory. None of the methods described above for the production of synthetic heliotropin from pyrocatechol can compete with the classical isosafrol degradation procedure. However, a recently developed procedure [13] should pave the way for a radical change such that pyrocatechol *will* become a viable starting material for the large scale synthesis of heliotropin.

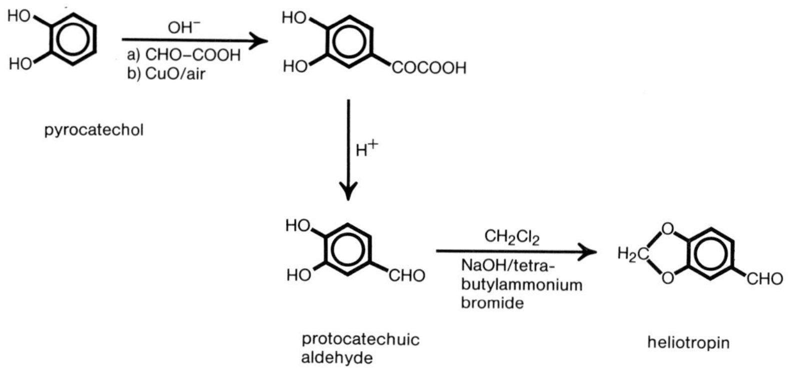

Fig. 5 Conversion of pyrocatechol to heliotropin via protocatechuic aldehyde.

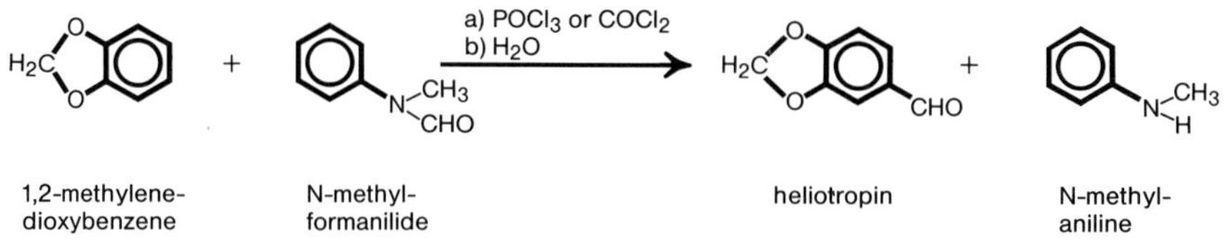

Fig. 6 Application of the Vilsmeier aldehyde synthesis to the production of heliotropin.

A new Synthesis of Heliotropin

In the course of our research, we investigated various methods for the synthesis of 3,4-methylenedioxymandelic acid, which can be considered a key intermediate in the synthesis of heliotropin. Unfortunately, the synthesis of this mandelic acid derivative, via the addition of glyoxylic acid to 1,2-methylenedioxybenzene under alkaline conditions, is not possible because of the absence of a phenolic hydroxyl group. The reaction of alkyl glyoxylates with aromatic hydrocarbons and their derivatives in acidic media has been studied by Werber [14], who showed that aromatic compounds with halogen or alkoxyl substituents give rise to diphenylacetic acids. Although the behavior of glyoxylic acid toward aromatics was not included in this study, there was no reason to believe that it would react any differently than alkyl glyoxylates. Thus, the chances of preventing the formation of diphenyl acetic acid derivatives in the present case were considered very slight, particularly in the presence of the reactive co-substrate phenol ether. Unexpectedly, we showed that the mandelic acid derivative can, in fact, be easily synthesized in excellent yield when 1,2-methylenedioxybenzene is reacted with glyoxylic acid in a strongly acidic medium at low temperature (Fig. 7).

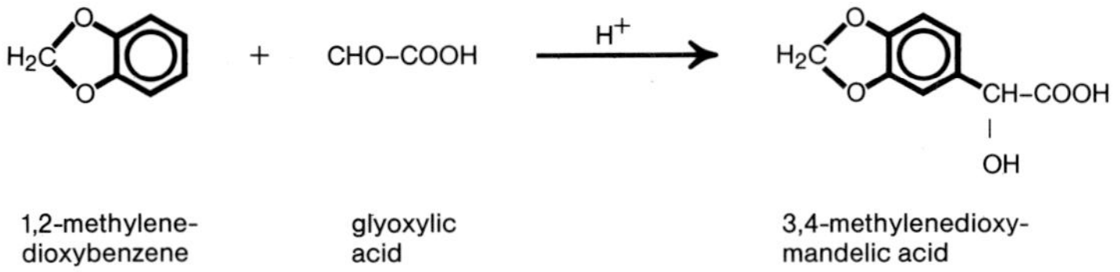

Fig. 7 Condensation of 1,2-methylenedioxybenzene and glyoxylic acid to 3,4-methylenedioxymandelic acid.

Glyoxylic acid, in the form of the monohydrate, is sufficiently acidic to catalyze the reaction. However, because of the 60–80 °C reaction temperature, considerable amounts of the disubstituted derivative, bis(3,4-methylenedioxyphenyl)acetic acid, are formed. This by-product is derived by the addition of a further molecule of 1,2-methylenedioxybenzene to the required product, 3,4-methylenedioxymandelic acid. With the addition of an acid catalyst stronger than glyoxylic acid (H_2SO_4, HCl or H_3PO_4), the reaction temperature can be lowered to the region of 5–20 °C with concomitant reduction in the formation of the diphenylacetic acid by-product. The desired product, 3,4-methylenedioxymandelic acid, crystallizes out of solution and can be separated by centrifugation for the subsequent oxidation step.

The oxidative decarboxylation of 3,4-methylenedioxymandelic acid has been investigated by Raman [15] who found that stoichiometric quantities of a silver iodide dibenzoate complex promoted decarboxylation, thereby yielding heliotropin. Surprisingly, we discovered that dilute nitric acid was an excellent oxidizing agent for the oxidative decarboxylation [16] (Fig. 8). The reaction proceeds very rapidly and almost quantitatively, even under mild conditions. The heliotropin which separates out either as an oil or as a crystalline mass, depending on the temperature, is then recovered by solvent extraction of the reaction mixture. It is interesting to note that the decarboxylation is not successful if the mandelic acid is oxidized to the corresponding phenylglyoxylic acid prior to treatment with nitric acid. This observation supports a postulated mechanism implicating the participation of a nitric acid ester (Fig. 8).

The scaling up of this new heliotropin synthesis to pilot plant dimensions was not entirely easy in that, toward the end of the first step in the synthesis, a crystalline mass of the mandelic acid is formed which is very difficult to stir. The crystallization is largely due to the low temperature which is necessary to prevent the formation of the diphenylacetic acid by-product.

The addition of solvent at this point, preferably acetic acid or water, has a positive effect on the course of the reaction and maintains the stirability of the mixture. The flow chart in Figure 9 illustrates the suitability of the process for continuous operation. Thus, the screw reactor is charged with methylenedioxybenzene, 50 % glyoxylic acid and 90 % sulfuric acid. After the

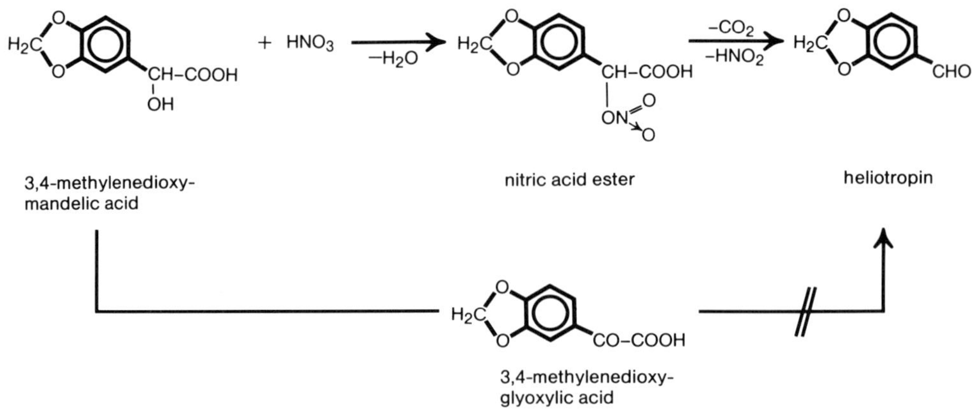

Fig. 8 Oxidative decarboxylation of 3,4-methylenedioxymandelic acid with dilute nitric acid.

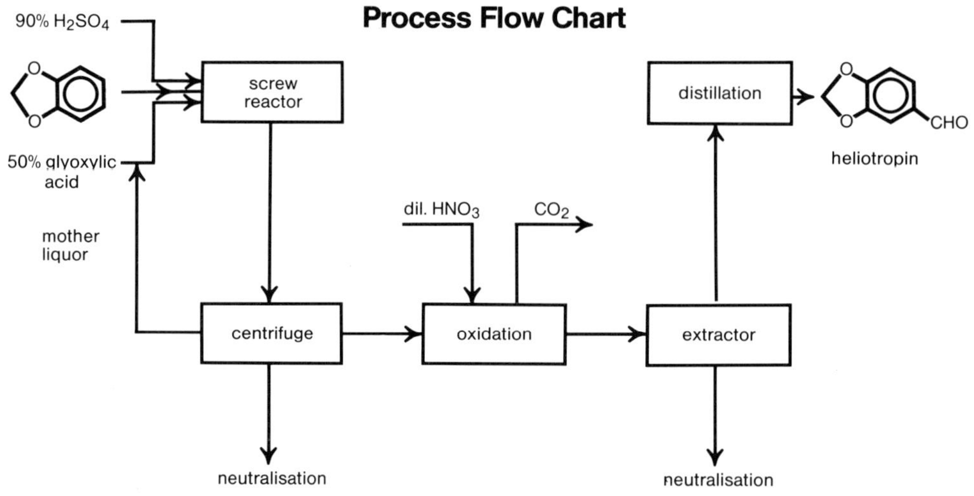

Fig. 9 Process flow chart for the continuous production of heliotropin.

reaction has gone to completion, the reaction mixture is centrifuged and the desired product, 3,4-methylenedioxymandelic acid, is collected and oxidized with dilute nitric acid to yield heliotropin. The crude heliotropin can be purified by distillation or by crystallization.

Due to the excellent overall yield of 80 %, this procedure for the total synthesis of heliotropin should be able to compete favorably with the semi-synthetic product obtained from safrol. Heliotropin is becoming increasingly important as a raw material for the synthesis of alkaloids, as described in paper No. 6 of this symposium. Furthermore, because heliotropin is still one of the most attractive synthetic fragrances, an even more interesting use is the incorporation of heliotropin into perfumes to obtain a certain exotic note.

References

[1] Arctander, S., Perfume and Flavor Chemicals, Montclair, N. J. (U.S.A.) (1969).
[2] Fittig, R., W. H. Mielch, Liebigs Ann. Chem. **153**, 25 (1869).
[3] Gildemeister, E., Die ätherischen Öle, Akademie-Verlag Berlin (1961).
[4] Eykman, J. E., Rec. Trav. chim. **4**, 32 (1885), Ber. **23**, 855 (1890).
[5] Ciamician, G., P. Silber, Ber. **23**, 1159 (1890).
[6] Schmitt, F., Parf. Kosm. **34**, 445 (1953).
[7] Lasskina, J. D., T. A. Dewitzkaja, P. F. Schilina, T. W. Ssucharukowa, S. N. Bytschkowa, U.S.S.R. 142302 (1961), Chem. Abstr. **57**, 7176c (1962).
[8] Jelissejeiva, W. N., T. A. Dewitzkaja, Angew. Chem. (russ.) **29**, 1894–96 (1958), Chem. Abstr. **51**, 7326d (1957).
[9] Bidlo, Z., F. Vonasek, Prumysla Potravin **9**, 589–92 (1958), C. **1961**, 15225.
[10] Umemura, S., N. Takamitsu, T. Enomiya, H. Shiraishi, T. Nakamura, Y. Ube to Ube Industries Ltd., DOS 2804063 (1978).
[11] Maggioni, P., to Brichima S.p.A. Mailand, DOS 2703640 (1977).
[12] Nakatani, K., to Mitsui Toatsu Chem. Ind., Japan Kokai 77156867.
[13] Bauer, K., R. Mölleken, to Haarmann & Reimer GmbH, Holzminden, U.S. 4190583 (1980).
[14] Werber, G., Gazz. Chim. Ital. **96** (4), 465 (1966), Chem. Abstr. **86**, 140020g (1977).
[15] Raman, P. S., Current Sci. (India) **27**, 22 (1958).
[16] Haarmann & Reimer GmbH, Holzminden, Research Lab., unpublished.
[17] Imoto, M., J. Soc. Chem. Ind. Japan **37**, 26B–29B (1934), C. **1934**, I, 1973.
[18] Milas, N. A., to Research Corp., U.S. 2414385 (1947), Chem. Abstr. **41**, 5895a (1947).
[19] Baranova, N. M., A. D. Yakusheva, P. N. Komarov, Maslo-Zhir.-Prom. **1974**, (5) 30–34, Chem. Abstr. **81**, 13219 (1974).
[20] Farinacci, N. T., U.S. 2794813 (1957), Chem. Abstr. **51**, 16572g (1957).
[21] Bogatskii, A. V., A. P. Antonov, G. L. Kamalov, Yu. V. Gardevich, V. V. Titov, I. A. Starovoit, E. G. Skestopalova, I. V. Smolovik, V. Ya. Kalashinkov, U.S.S.R. 490793 (1975), Chem. Abstr. **84**, 74254s (1976).
[22] Interchemical Corp., Brit. 1092615 (1967), Chem. Abstr. **69**, 2516w (1968).

6. AN ECONOMICAL SYNTHESIS OF PEPPER ALKALOIDS

H. J. Oediger* and A. Schulze*

Summary Pepper alkaloids of the general formula

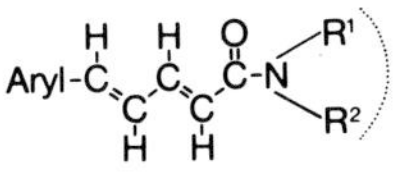

can easily be synthesized from the corresponding aldehydes and the corresponding amides of crotonic acid. Thus, the pungent principle of black pepper, piperine (*1*), is obtained through base-catalyzed condensation of piperonal (*2*) and 1-crotonoyl-piperidine (*3*).

Scheme 1

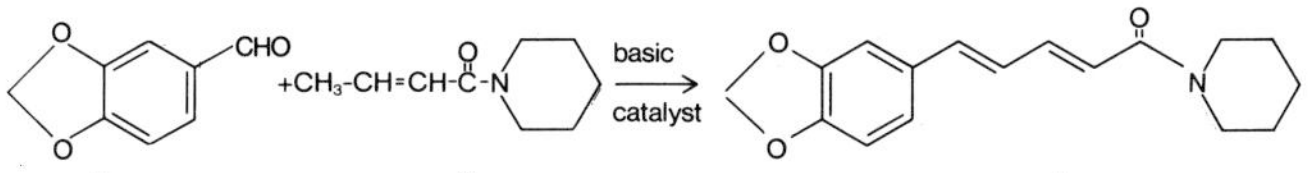

The scope and limitations of the reaction are described.

Introduction

Pepper, noted for its pungent taste, is one of the most important condiments in the world. Pepper plants belong to two different botanical families, the solanaceae and the piperaceae. Capsicum (e.g., the paprika group) is known as "red pepper" and is a member of the solanaceae, closely related to the potato, tomato and tobacco. This group will not be discussed here. Pepper belongs to the piperaceae family, and *Piper nigrum*, which is known as "black pepper" or "white pepper" depending upon the way it has been treated, is the dominant member of this group. The plant is cultivated in Java, Sumatra, India, Ceylon and other areas in this region of the world. *Piper nigrum* is a creeper and its grape-like fruits are green when unripe, and turn red upon ripening [1].

* H. J. Oediger and A. Schulze, Bayer AG, 5090 Leverkusen, West Germany.

Composition of *Piper nigrum* and Related Species

Piper nigrum is not only characterized by its extremely pungent taste, but also by its wealth of volatile constituents found in the so-called "pepper oil". The oxygen-containing components of the oil have been recently re-investigated by Debrauwere and Verzele [2], who succeeded in identifying some 33 compounds. Among these (Fig. 1) were piperonal (heliotropin) (*2*) and

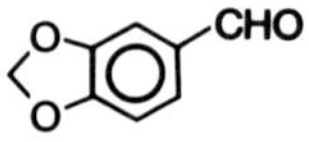

(2)

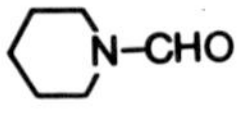

(4)

Fig. 1 Some volatile components of pepper oil.

1-formyl piperidine (*4*), which could be metabolically related to piperine (*1*). *Piper nigrum* contains, in addition to piperine (*1*), which is characterized by its *E,E*-configuration, several other compounds [3] which are closely related to piperine, including piperyline (*5*) and the two piperoleines, A and B (*6*) (Fig. 2). Piperyline (*5*) differs structurally from piperine in that it has a pyrrolidine ring instead of a piperidine ring; yet, it also has a pungent taste. On the other hand, the piperoleines A and B have different side-chains compared to piperine.

There are other *Piper* species in addition to *Piper nigrum* that have pungent tastes, and in each case the pungent principle is chemically very similar to piperine. *Piper guineense*, for example, is cultivated in West Africa, and from it both wisanine (*7*) and okolasine (*8*) (Fig. 2) have been isolated [4]. The chemical difference compared to piperine (*1*) and piperyline (*5*) of *Piper nigrum* is the additional methoxy-group on the aromatic ring (Fig. 2). Both wisanin and okolasin were synthesized by Scharf *et. al.* [5] via the phosphonium salt (*10*) and the recently discovered ester of fumaric aldehyde (*9*) as illustrated in Figure 3. Isomers of wisanine have also been synthesized [6].

The pungent principle of *Piper trichostachyon*, which is known as trichostachine [7], is, in fact, identical to piperyline (*5*) found in *Piper nigrum* (Fig. 2). Additionally, a second alkaloid (*11*) was isolated from this species [8] and shown to be a higher homologue of trichostachine (i.e., containing an additional ethene grouping) (Fig. 2). The synthesis of (*11*) was achieved by means of two Wittig-Horner reactions using piperonal (*2*), piperonylidene acetaldehyde and the ethyl ester of bromocrotonic acid as starting materials [9]. *Piper longum* grown in India contains, in addition to piperine, both piperlonguminine (*12*) with its characteristic isobutylamine group, and piplartine (*13*) which displays a rather different chemical structure [10] (Fig. 2).

Syntheses of Piperine

Considerable emphasis has been placed on the development of a commercial synthesis of piperine, the pungent principle of *Piper nigrum*. As indicated previously, the natural product has the *E,E*-structure (Fig. 2). On exposure to light, however, solutions of piperine undergo photoinduced isomerization to yield an equilibrium mixture of the four possible geometric isomers [11] (Fig. 4). This fact is very important for all synthetic aproaches to piperine, since

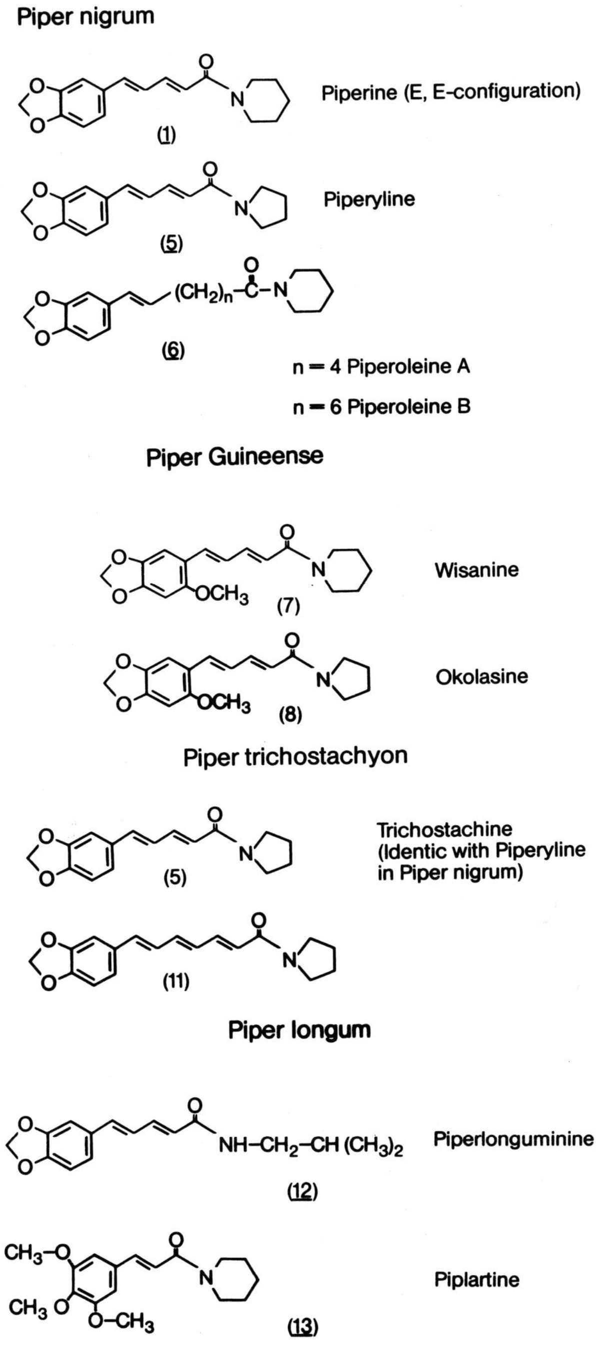

Fig. 2 Pungent constituents of *Piper* species.

only the *E,E*-isomer (*17*) is of interest. Logical fragmentation of piperine at locations A and B (Fig. 5) suggests two feasible schemes for the synthesis of this compound, namely the $C_3 + C_2$ approach and the $C_1 + C_4$ approach, where the subscripts indicate the number of carbon atoms in the side-chain synthons. In both approaches, bond formation between the synthons can be achieved by the Wittig or Horner reaction with the aid of organophosphorus intermediates.

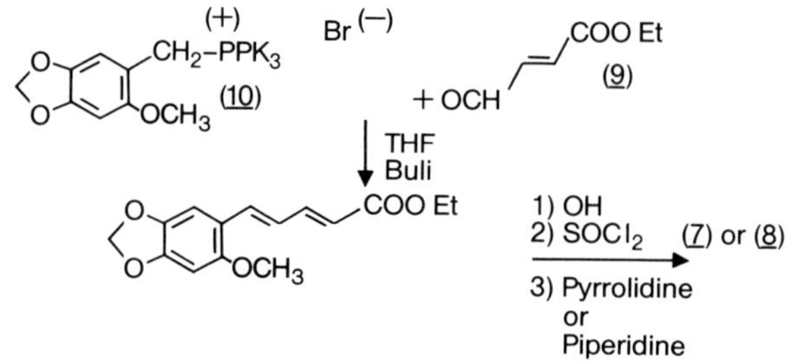

Fig. 3 Synthesis of wisanin (*7*) and okolasin (*8*) by the procedure of Scharf *et al.* [5].

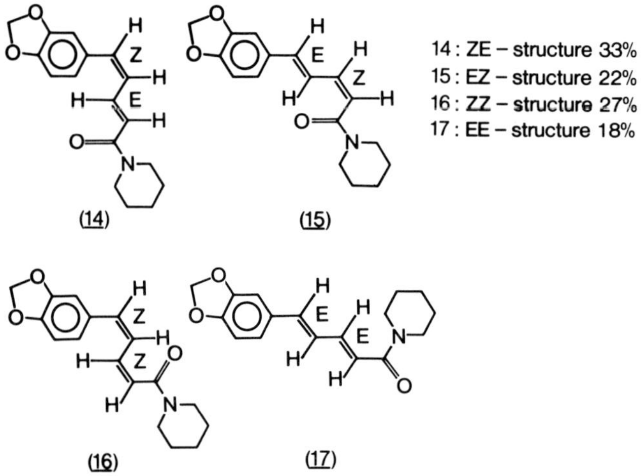

Fig. 4 Photo-equilibrated isomer composition of piperine (*17*).

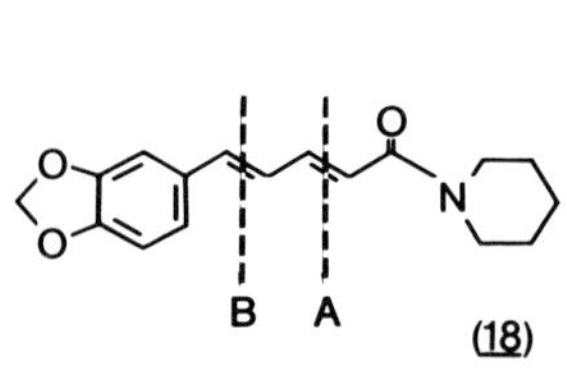

Fig. 5 Alternatives for the synthesis of piperine by the $C_3 + C_2$ approach (cleavage A) and the $C_1 + C_4$ approach (cleavage B).

The C_3 + C_2 Approach

One example of the $C_3 + C_2$ approach is illustrated in Figure 6, whereby the aldehyde (*20*) (C_3-unit) is reacted with the phosphonate (*21*) (C_2-unit) to form piperine (*1*). The phosphonate (*21*) (C_2-unit) can be synthesized using well-known synthetic methods (Fig. 7). Thus, chloroacetylchloride (*23*) is treated with piperidine and the resulting chloroacetylpiperidine (*22*) affords the phosphonate (*21*) in excellent yield after reaction with triethylphosphite (*24*). The essential C_3-unit, aldehyde (*20*), can be prepared by direct condensation of piperonal (*2*) and acetaldehyde, but the yield is low [12]. For the synthesis of the aldehyde (*20*), we therefore chose a general procedure (Fig. 6) similar to that published by Dallacker [13]. Thus, starting with the

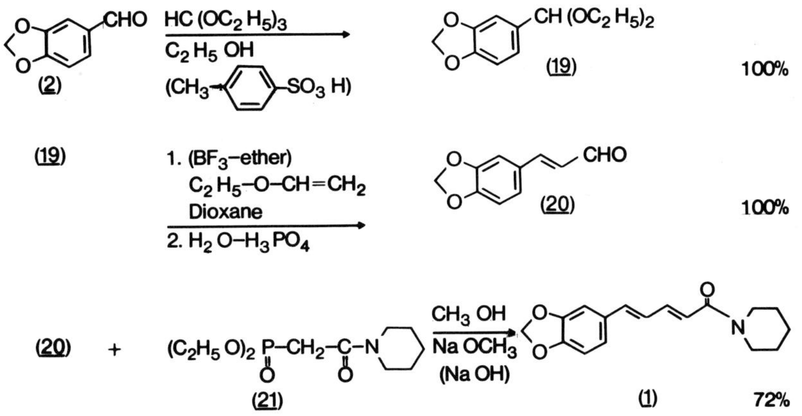

Fig. 6 A procedure for the synthesis of piperine by the $C_3 + C_2$ approach.

acetal (*19*) obtained from piperonal (*2*), the aldehyde (*20*) is prepared by enol ethercondensation with ethyl vinyl ether. Using this procedure, high yields are obtained only when the intermediates are not purified, and when the acetal (*19*) and ethyl vinyl ether are condensed by treatment with BF_3-ether instead of $ZnCl_2$. Although piperine can be obtained in 72 % yield, this synthesis is not of commercial interest because there are too many steps.

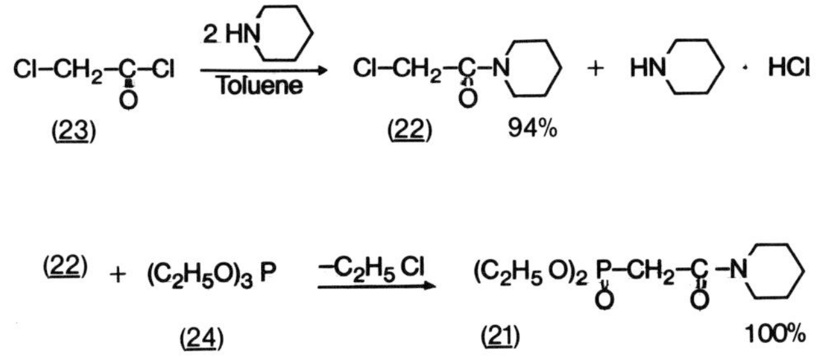

Fig. 7 Procedure for the synthesis of the phosphonate C_2-unit.

Recently, Tsuboi and co-workers [14] published a piperine synthesis in which the $C_3 + C_2$ strategy was also used. In this procedure (Fig. 8), piperonal (*2*) is reacted with acetylene to form the carbinol (C_3-unit) (*25*), which can be condensed with the acetal (C_2-unit) (*26*) to form compound (*27*) containing an allene group. The last step in this synthesis is the isomerization of (*27*) to form a mixture of *E,E*- and *E,Z*-structures in a 65:35 ratio.

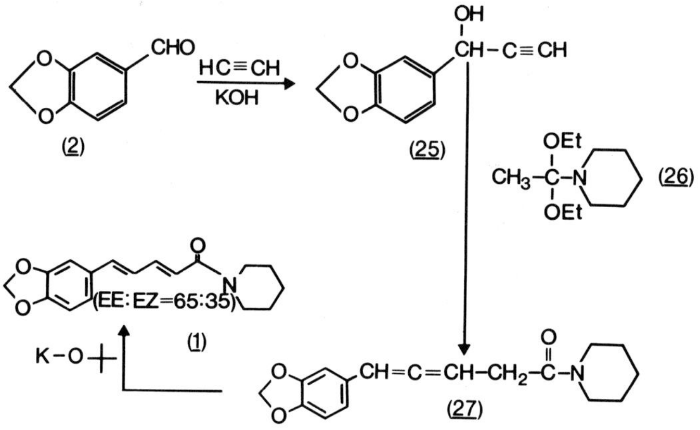

Fig. 8 Synthesis of piperine (*1*) by the $C_3 + C_2$ approach of Tsuboi *et al.* [14].

The $C_1 + C_4$ Approach

Although many different types of reactions are available for joining the C_1-unit to the C_4-unit, we initially explored only the two possibilities illustrated in Figure 9. Thus, the so-called C_1-unit aldehyde (*2*) can react with the C_4-unit halide (*28*), or the C_1-unit halide (*29*) can react with the C_4-unit aldehyde (*30*) using the Wittig-Horner reaction. However, the C_4-units (*28*) and (*30*), are not readily available and their syntheses require several steps.

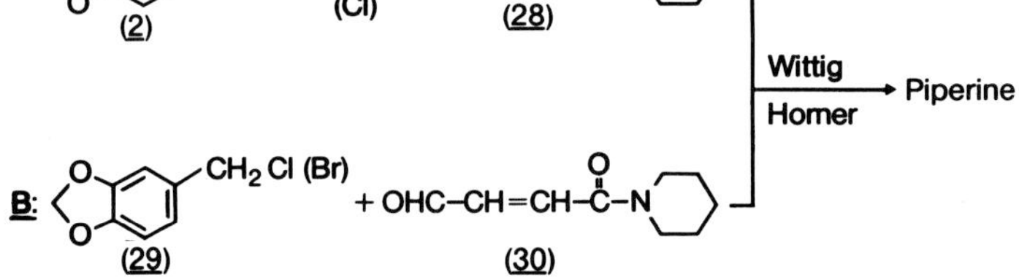

Fig. 9 Some $C_1 + C_4$ strategies for the synthesis of piperine.

Therefore we turned to another approach, as illustrated in Figure 10. Condensation of the hydroxylactone (*32*) [15], with the phosphonium salt (*31*) prepared from piperonal (*2*), gives the piperine acid (*33*), which can then be transformed into piperine (*1*). However, this synthetic approach is unsatisfactory because the yield of acid (*33*) is low.

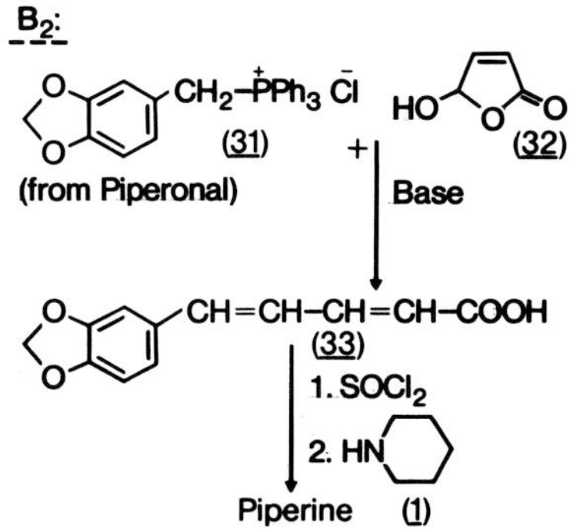

Fig. 10 An alternate strategy for the synthesis of piperine.

An Economical Synthesis of Piperine and Related Alkaloids

Two functional groups are necessary in order to join together the carbon units via the synthetic reactions described above. This fact necessarily makes these syntheses expensive. Consequently, we searched for a simple procedure which would enable the joining together of carbon atoms by using only one functional group. The aldol condensation is an example of such a reaction, and we therefore explored this synthetic strategy. After numerous experiments with a variety of basic reagents, which provided piperine yields in the 50–60 % range, we found [16] that the system indicated in Figure 11 would increase the yield to about 70 %. In this procedure piperonal (*2*) reacts with crotonoylpiperidine (*3*), in dimethylsulfoxide under the catalytic

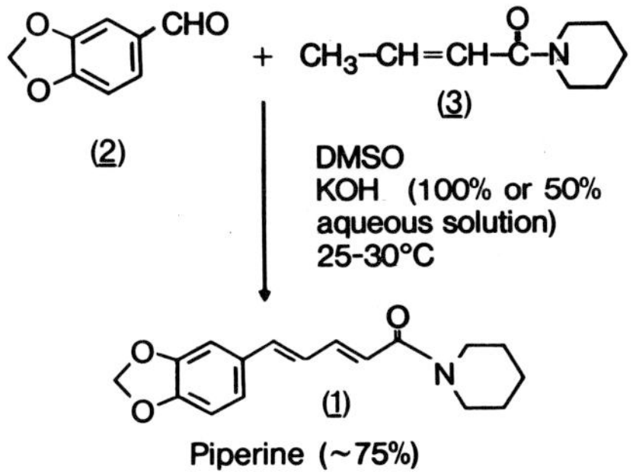

Fig. 11 Condensation of piperonal (*2*) with crotonoylpiperidine (*3*) to yield piperine (*1*).

influence of potassium hydroxide, to yield piperine (*1*) in good yield. The crotonoylpiperidine (*3*) may be prepared from crotonic acid (*34*) in several ways (Fig. 12). Thus, for example, crotonic acid (*34*) can either be treated with thionyl chloride and the resulting acid chloride reacted with piperidine in the presence of base, or the acid (*34*) can be condensed directly with piperidine in boiling toluene.

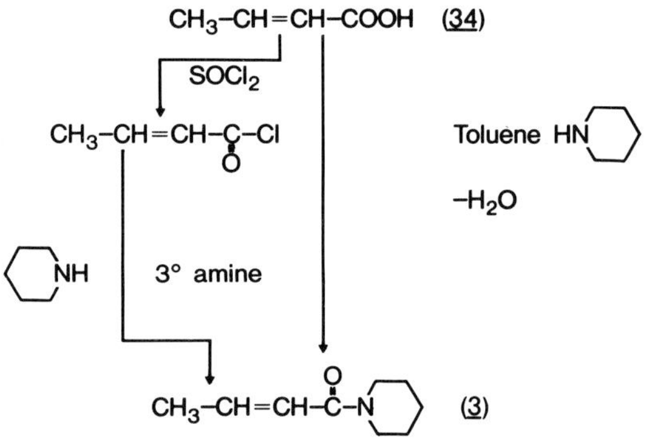

Fig. 12 Procedures for the preparation of crotonoylpiperidine (*3*) from crotonic acid (*34*).

While utilizing the basic reaction scheme in Figure 11, we attempted to improve the process by increasing the yield to 80 % or more, by decreasing the reaction time, and by substituting a more common solvent for dimethylsulfoxide.

Phase-transfer reactions are now very common, but only a few examples [17a, 17b] exist for the aldol condensation-type of reaction. In such an application the reaction partners are found in the organic phase, while the aqueous phase contains concentrated sodium hydroxide and a quaternary ammonium salt. The latter carries (transfers) the hydroxide ion into the organic phase and thereby initiates the aldol condensation reaction. Thus, the usual rules for two-phase reactions do not apply to the preparation of piperine by this procedure. The system $NaOH/CH_2Cl_2$ is unsuccessful; yet, good yields (75–80 %) are obtained by the substitution of 50 % aqueous potassium hydroxide for sodium hydroxide, and by the use of a small quantity of dimethylsulfoxide (or in some cases without any solvent at all). However, the use of large quantities of 50 % aqueous potassium hydroxide is not feasible for the large-scale production of piperine. By systematic study, we demonstrated that the quantity of potassium hydroxide could be reduced to a level equivalent to that of the phase-transfer catalyst with excellent results (at 60–65 °C). Table 1 shows the effect of the solvent on the yield of piperine under these standard reaction conditions, using triethylbenzylammonium chloride (TEBA) as the transfer catalyst. While a variety of phase-transfer catalysts can be successfully employed, quaternary ammonium hydroxides are especially effective [18]. Table 2 illustrates the effect of such quaternary compounds on the phase-transfer reaction used for the synthesis of piperine under standard conditions. By combination of the appropriate solvent and ammonium compound (eg., ethyleneglycoldimethyl ether and benzyldimethylphenylammonium hydroxide) the yield of piperine is greater than 80 %. Similar yields can also be obtained by the use of crown ethers and alkaline metal hydroxides, such as the combination 18-crown-6 and 50 % aqueous potassium hydroxide [18].

Table 1. Effect of different solvents on the yield of piperine under standard reaction conditions.

solvent	Piperine % yield
Methyl sulfoxide	80
Anisole	77
Ethylene glycol dimethyl ether	75
N,N–Dimethylacetamide	72
Diethylen glycol dimethyl ether	72
N,N–Dimethyl benzylamine	70
1–Methyl–2–pyrrolidinone	67
tert.-Butanol	65
Dioxane	62
Sulfolane	62
Tetrahydrofuran	59
N,N–Dimethylformamide (Na OH) 46 / (KOH) 40	
Ethanol	11
Acetonitrile	5
Dichloromethane	~5
Ethyl acetate	–
———	64

Table 2. Effect of different quaternary ammonium hydroxide phase-transfer catalysts on the yield of piperine.

$[R_4N]OH$	Piperine % yield
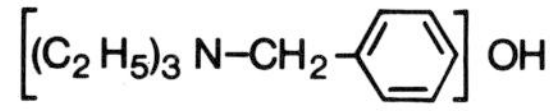	80
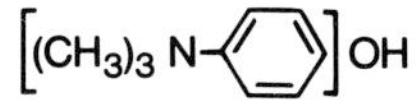	75
$[(C_2H_5)_4N]OH$	71
$[(C_4H_9)_4N]OH$	75
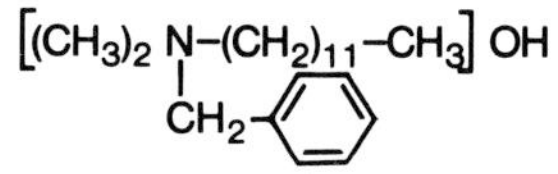 50% aqueous solution	74
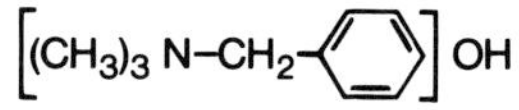 40% aqueous solution	75

Thus, by using simple intermediates (piperonal and crotonoylpiperidine) and appropriate condensation reaction parameters, piperine can be obtained in good space/time yields in normal technical systems. Other pepper constituents, such as piperyline, wisanine, okolasine, piperlonguminine and piplartine, can be obtained using the same general synthetic approach.

Acknowledgement

We would like tho thank Drs. Karl, Niedrig, Ritter and Wendisch, Central Research and Development, Bayer AG, for carrying out the analytical aspects of this work.

References

[1] Newman, A. A., Chemical Products, 343 (1953).
[2] Debrauwere, J., and M. Verzele, Bull. Soc. Chim. Belg. **84**, 167 (1975).
[3] Grewe, R., W. Freist, H. Neumann and S. Kersten, Chem. Ber. **103**, 3752 (1970).
[4] Addac-Mensah, I., and F. G. Torto, Tetrahedron Lett., 3049 (1976).
[5] Scharf, H. D., J. Janus, F. Dallacker and R. Morcinek, Justus Liebigs Ann. Chem., 573 (1978).
[6] Scharf, H. D., and J. Janus, Tetrahedron **35**, 385 (1979).
[7] Singh, J., K. L. Dhar and C. K. Atal, Tetrahedron Lett., 4975 (1969).
[8] Singh, J., and M. A. Potdar, Phytochemistry **13**, 677 (1974).
[9] Vig, O. P., V. D. Ahuja, B. Vig and A. K. Vig, Ind. J. Chem. **13**, 1015 (1975).
[10] Chatterjee, A., and C. P. Dutta, Tetrahedron, 1769 (1967).
[11] De Cleyn, R., and M. Verzele, Chromatographia **8**, 342 (1975).
[12] Poirier, Y., and N. Lozac'H, Bull. Soc. Chim. France, 865 (1967).
[13] Dallacker, F., and J. Schubert, Chem. Ber. **108**, 95 (1975).
[14] Tsuboi, S., and A. Takeda, Tetrahedron Lett., 1043 (1979).
[15] Deutsches Bundespatent 2111119, BASF (Erf.: G. Bolz and W. W. Wiersdorff).
[16] Deutsches Bundespatent 2757506, Haarmann & Reimer (Erf.: A. Schulze and H. Oediger).
[17a] Cardillo, G., D. Savoia and A. Umani-Ronchi, Synthesis, 453 (1975).
[17b] Dryanska, V., and C. Ivanov, Tetrahedron Lett., 3519 (1975).
[18] Deutsches Bundespatent 2757483, Haarmann & Reimer (Erf.: H. Oediger and A. Schulze).

7. A TWO-STEP PROCEDURE FOR THE SYNTHESIS OF ANETHOLE

R. G. Mölleken* and K. A. Bauer*

Summary A short review of known methods for the synthesis and production of anethole (*1*) is followed by a detailed description of a new synthetic procedure using anisole and propionaldehyde as starting materials. Condensation of anisole (*2*) with propionaldehyde (*4*) gives the bisphenol ether (*3*) which undergoes thermal cleavage in the presence of an acidic catalyst to yield anethole (*1*) and anisole (*2*).

Scheme 1

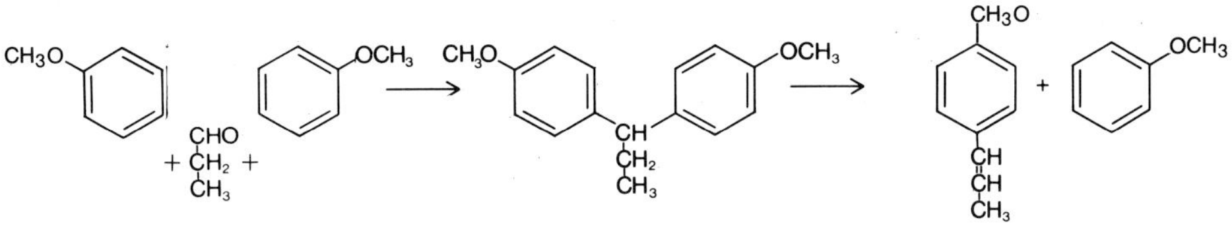

(2) (4) (2) (3) (1) (2)

Problems experienced in the cleavage of the bisphenol ether (*3*) under industrial conditions and in the isolation of anethole are described. Areas of application for the synthetic anethole are also discussed.

Introduction

The systematic name for anethole is 1-methoxy-4-(1-propen-1-yl-)benzene, and this compound can exist in both the *cis*- and *trans*-form as shown in Figure 1. The economically more important of the two isomers is *trans*-anethole, the main natural sources of which are star aniseed oil [1] and fennel oil [2, 3]. Due to the extremely high levels found in these oils, a simple freezing-out technique can be employed to obtain pure *trans*-anethole. The lower levels of anethole and estragole [4, 5] present in pine oil and in sulphate turpentine processing wastes are, however, of greater importance technologically. (Estragole is the trivial name for *p*-allyl anisole,

* R. G. Mölleken and K. A. Bauer, Haarmann & Reimer GmbH, 3450 Holzminden, West Germany.

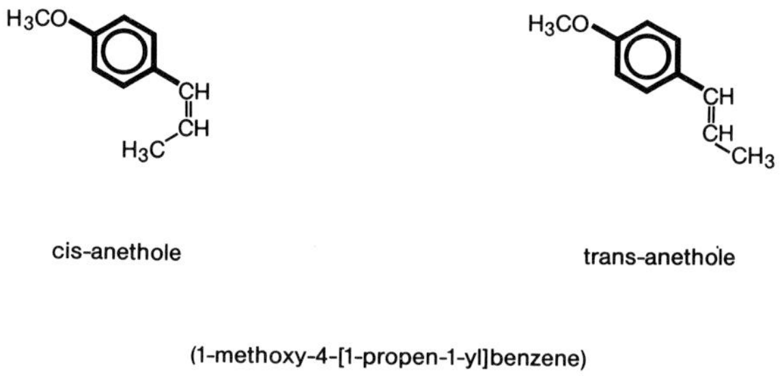

Fig. 1 Geometric isomers of anethole (1-methoxy-4-(propen-1-yl)benzene).

and it is readily isomerized to yield anethole, Fig. 2.) Sulphate turpentine is a by-product of the kraft sulphate pulping process for the production of paper, and in 1976 some 20 million gallons were produced in the U.S. alone [4]. The combined anethole and estragole contents of pine oil amount to approximately 1.5 %, so it is readily apparent that 1000 tons of anethole could be obtained annually from this source alone. Therefore, it is not surprising that very few new processes for the manufacture of fully synthetic anethole can be found in the literature [6, 7, 8].

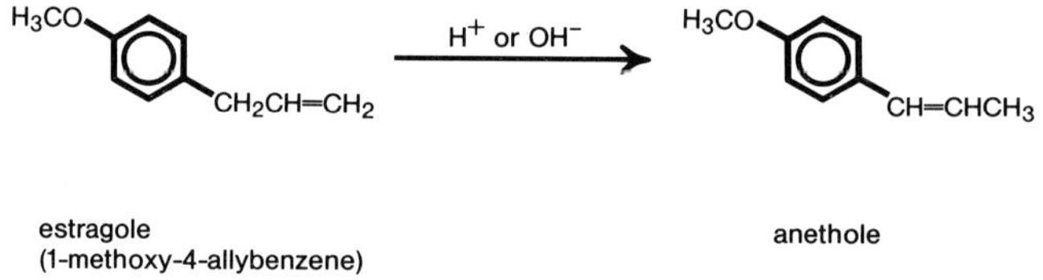

Fig. 2 Isomerization of estragole to anethole.

Review of Anethole Syntheses

All of the technically feasible processes for the preparation of anethole use anisole as starting material. Quelet [9] allowed anisole and propionaldehyde to react together in the presence of hydrogen chloride gas and phosphoric acid. The product of this reaction was the corresponding α-chloropropyl compound, which on elimination with pyridine yielded anethole (Fig. 3). The overall yield of this process is relatively low (25 %). In another process, developed by Russian investigators [10], anisole was reacted with the acid chloride or anhydride of propionic acid, thereby obtaining *p*-methoxypropiophenone (Fig. 4). This compound was subsequently reduced to the corresponding secondary alcohol using copper chromite and hydrogen, and the alcohol then dehydrated to anethole with the assistance of acid catalysts. The overall yield of this process, which is probably being industrially realized in the Soviet Union, is approximately 50 %. An alternative synthetic process involving estragole [11] (Fig. 5), which is easily isomerized to anethole under both acidic or basic conditions [12], does not appear to be industrially feasible because of the extremely low overall yield.

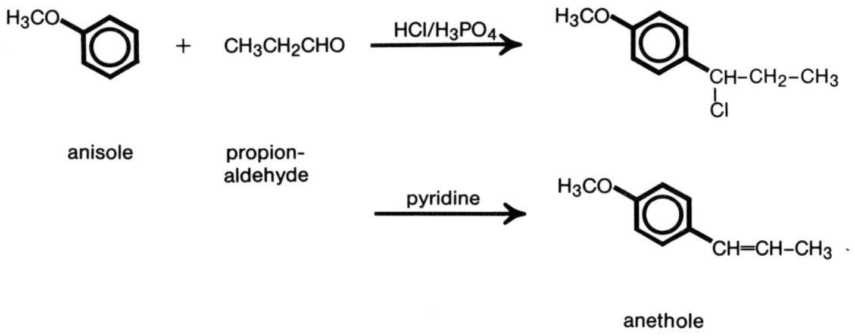

Fig. 3 Synthesis of anethole by the method of Quelet [9].

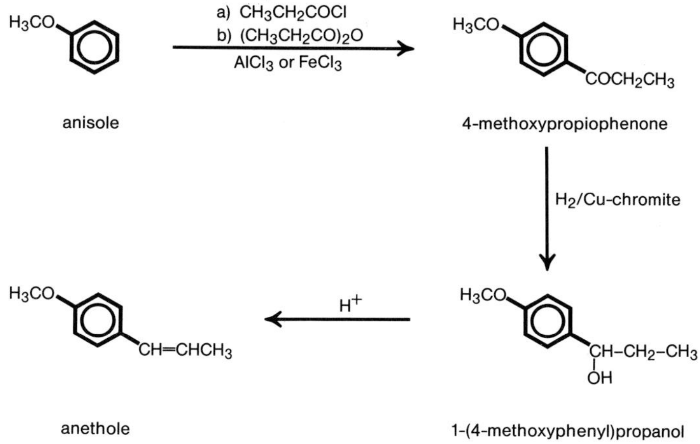

Fig. 4 Synthesis of anethole by the method of Svadkovskaya *et al.* [10].

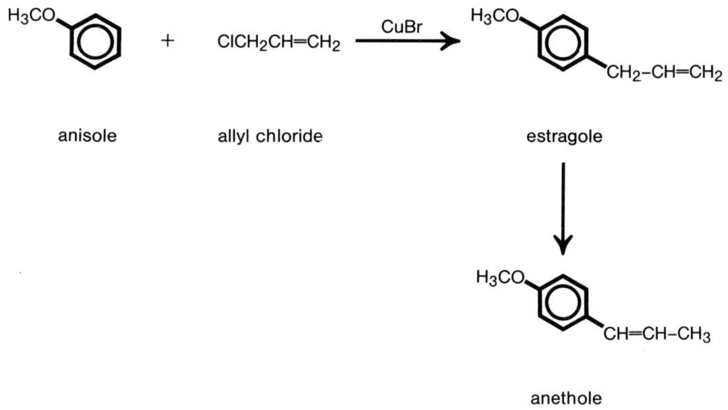

Fig. 5 A synthesis of anethole via estragole.

A Two-Step Procedure for the Synthesis of Anethole

The economic importance of anethole, which is used very widely in the food industry, prompted us to search for a less complicated method for the synthesis of this compound. During the course of extensive investigations into the reaction between aldehydes and phenol derivatives (described in paper No. 3 of this symposium), it was found that anisole could be converted into a mixture of 1,1-bis(methoxyphenyl)propanes in the presence of acid catalysts [13]. The main product of the mixture is 1,1-bis(4-methoxyphenyl)propane (Fig. 6). Phosphoric

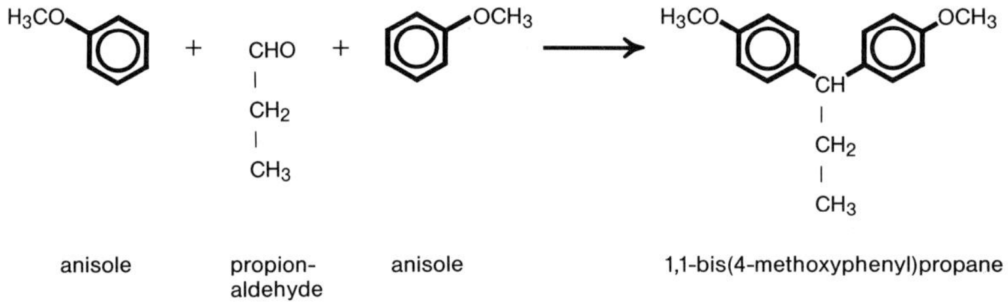

Fig. 6 Condensation of anisole and propionaldehyde to 1,1-bis(methoxyphenyl)propane.

acid is an especially suitable catalyst, as are inorganic ion exchangers based upon aluminium silicate. The condensation reaction is best carried out by mixing the anisole with the catalyst, warming this mixture in the reaction vessel to 140–150 °C under atmospheric pressure, and then adding propionaldehyde to the mixture at a rate determined by the speed of the condensation reaction. This technique guarantees that the water, which is formed as one of the reaction products, can be continually removed from the reaction mixture, thereby ensuring that the reaction goes more-or-less to completion. It was found that water rather than propionaldehyde distills over under these conditions, and this was surprising since propionaldehyde has a much lower boiling point (46.5 °C) than does water. Additionally, this technique drastically reduces the amount of 2-methylpentenal formed by self-condensation of propionaldehyde. The reaction mixture consists of approximately 80 % *p,p'*- and 20 % *o,p'*-dimethoxydiphenyl propane. This mixture of dimethoxydiphenyl propanes and catalyst undergoes thermal cleavage to yield anethole and anisole when the temperature is raised to 180–200 °C (Fig. 7). The cleavage reaction is carried out *in vacuo* so that those products which boil below 180°C distill off immediately.

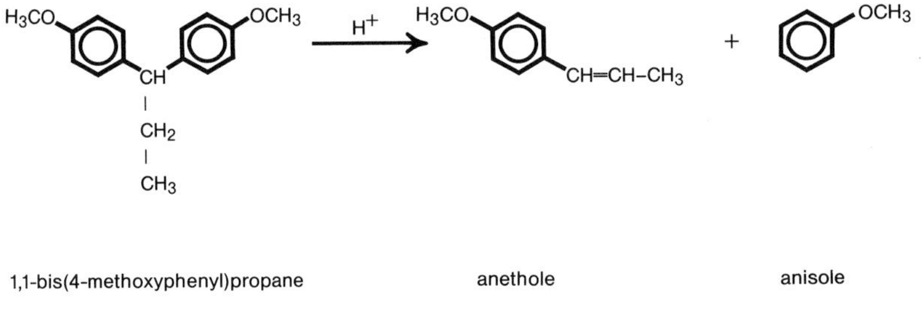

Fig. 7 Thermal cleavage of 1,1-bis(4-methoxyphenyl)propane to anethole and anisole.

The distillate contains anisole and the *cis*- and *trans-p*-anetholes, the latter two being present in the ratio of 15:85. In addition to these compounds, propyl anisole is also formed. If a crude mixture of the isomeric dimethoxydiphenyl propanes is used for the cleavage reaction, then *cis*- and *trans-o*-anethole will also be present. Fractional distillation can be used to remove these compounds from the *trans-p*-anethole. The *cis-p*-anethole formed in the reaction is either isomerized in a separate step to form *trans-p*-anethole [14], or it is fed back, along with the *cis*- and *trans-o*-anetholes, to the condensation step. Under the condensation reaction conditions, the *o*-anetholes react with anisole to form 1-(2-methoxyphenyl)-1-(4-methoxyphenyl)propane which can be cleaved to yield, at least theoretically, 50% *p*-anethole (Fig. 8). Although the

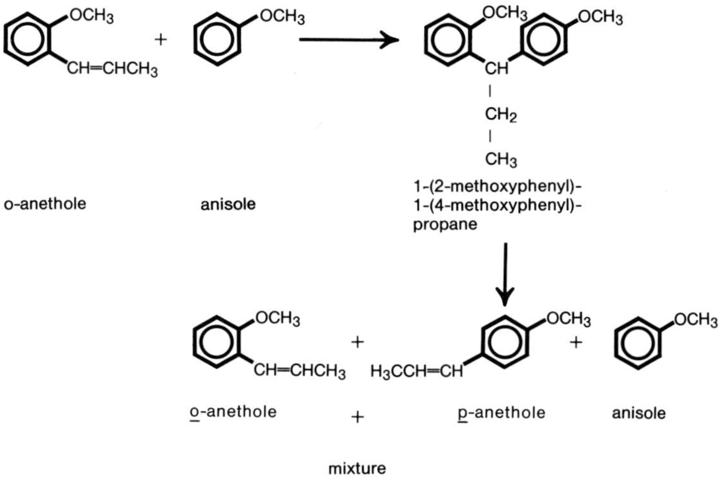

Fig. 8 Pathway for the recycling of *o*-anethole.

anethole isomers, which are by-products of the reaction, appear to have no technical use, they can be recycled in this manner. Thus, the whole process becomes more economically viable by keeping by-product formation to a minimum.

It is well-known that anethole is extremely unstable towards acids, and that even traces of acid can lead to dimerization and further polymerization. Consequently, in scaling up the cleavage step very difficult technical problems had to be overcome. From our experience, the best results are obtained when reaction vessels are used which ensure that the anethole evaporates immediately as it is formed, thereby reducing contact with the acid to an absolute minimum. Suitable reactors are thin film evaporators and molecular distillation units. The latter, which are especially suitable, are now available in a compact form with large capacity. The use of anisole as a carrier vapor in the cleavage step has also been very successful. A simple process flow scheme illustrating the production of *trans*-anethole is given in Figure 9. The final purification of the synthetic anethole is best achieved by crystallization. The very pure *trans*-anethole obtained by this procedure is indistinguishable from natural anethole with regard to the normal physical, chemical and organoleptic properties.

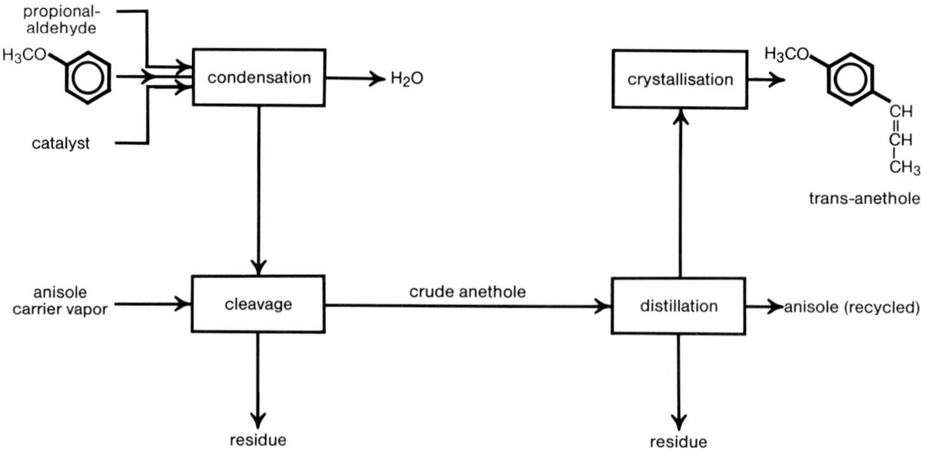

Fig. 9 Process flow chart for the production of *trans*-anethole.

This relatively simple synthesis of anethole should make a meaningful contribution toward satisfying a steadily increasing demand, which until now has outpaced supply. In addition, *trans*-anethole of constant quality and stable price will now become available on the market, and it is, therefore, expected that the use of this compound in foodstuffs and medicinals will increase without limitation.

References

[1] Cahours, A., C. R. Acad. Sci. (Paris) **12,** 1213 (1841); Liebigs Ann. Chem. **35,** 313 (1840).
[2] Cahours, A., Liebigs Ann. Chem. **41,** 74 (1842).
[3] Costa, A. F., J. C. Dovale, Perfum. Record **50,** 291 (1958).
[4] Derfer, J. M., Perfumer and Flavorist **3,** (1) 45 (1978).
[5] Garber, M., R. Herrlinger, H. Albrecht, L. F. Ciesielski, to American Cyanamid Co. U.S. 3038945 (1962).
Kirk Othmer, Encyclopedia of chemical technology, 2nd. ed., Wiley-Interscience, New York, Vol. **19,** 819–820.
[6] Naves, Y.-R., Riv. Ital. **56,** 630–634 (1974).
[7] Hirav, N., Kinki Daigaku Rikegakubu Kenkyu Hokoku **1977** (12), 71–76, Chem. Abstr. **88,** 89246x (1978).
[8] Kheifits, L. A., G. E. Svadkovskaya, A. J. Platova, 6th Intern. Congress Essent. Oils, **1974,** 133, Chem. Abstr. **84,** 30545h (1976).
[9] Quelet, R., C. R. Acad. Sci. (Paris) **202,** 956 (1936).
[10] Svadkovskaya, G. E., L. A. Kheifits, A. J. Platova, W. J. Denisenkowa U.S.S.R. 261380 (1968), Chem. Abstr. **73,** 14450r (1970).
[11] Melkanowitzkaya, Sc. G., J. P. Zukerwanik, Ber. Akad. Wiss. Usbek S.S.R. (1959), (No. 11), 40–44, C. **1963,** 19205.
[12] Metal and Thermit Corp. Fr. 1197634, C. **1962,** 7321.
[13] Bauer, K., R. Mölleken, Haarmann & Reimer GmbH, Holzminden, DAS 2418974 (1974).
[14] Platova, A. J., U.S.S.R. 455087, Chem. Abstr. **82,** 170360c (1975).

8. THE ADDITION OF ALDEHYDES TO ACTIVATED DOUBLE BONDS AND ITS APPLICATION IN THE SYNTHESIS OF JASMINE FRAGRANCES

H. Kuhlmann* and H. Stetter*

Summary Aromatic and heterocyclic aldehydes readily undergo addition to α,β-unsaturated ketones, esters and nitriles thereby yielding the corresponding 1,4-diketones, 4-oxocarboxylic acid esters and 4-oxonitriles, respectively. The reaction is catalyzed by cyanide ions and is performed under aprotic conditions. Thiazolium salts are also suitable catalysts when used in the presence of a base. Under these conditions, not only do aromatic and heterocyclic aldehydes undergo the addition reaction, but also aliphatic aldehydes. The reaction can be performed both with and without a solvent, and high yields are usually obtained. A large number of 1,4-diketones and γ-ketoacids, both with and without additional functional groups, have been synthesized. Aliphatic 1,4-diketones easily cyclize, via an aldol condensation, to form cyclopentenones. Consequently, this procedure respresents an efficient and elegant method for the synthesis of jasmine fragrances, including *cis*-jasmone, dihydrojasmone and methyl dihydrojasmonate.

Introduction: The Addition of Aldehydes to Activated Double Bonds

One of the classical reactions in organic chemistry is the benzoin reaction in which aromatic and heterocyclic aldehydes are transformed into α-hydroxyketones, the benzoins [1] (Fig. 1). Cyanide ions serve as catalysts for the reaction, which is reversible and proceeds via an intermediate carbanion that is stabilized by the nitrile group [2]. Stetter and Schreckenberg found that this carbanion can be added to the double bond of α,β-unsaturated ketones [3], esters [4] and nitriles [5] (Fig. 1), thus yielding γ-diketones, 4-oxocarboxylic acid esters, and 4-oxonitriles as the respective products (Fig. 2). The main drawback of this reaction is, however, its complete failure with aliphatic aldehydes [6]. When attempting this reaction with such aldehydes in the presence of cyanide ions, a vivid color development and tar formation is observed. The preferential and uncontrolled aldol condensations that cyanide ions promote with aldehydes are responsible for this color change. Thus, in order to carry out the addition with aliphatic aldehydes, it was necessary to develop more effective catalysts.

* H. Kuhlmann and H. Stetter, Department of Organic Chemistry, Technical University of Aachen, 5100 Aachen, West Germany.

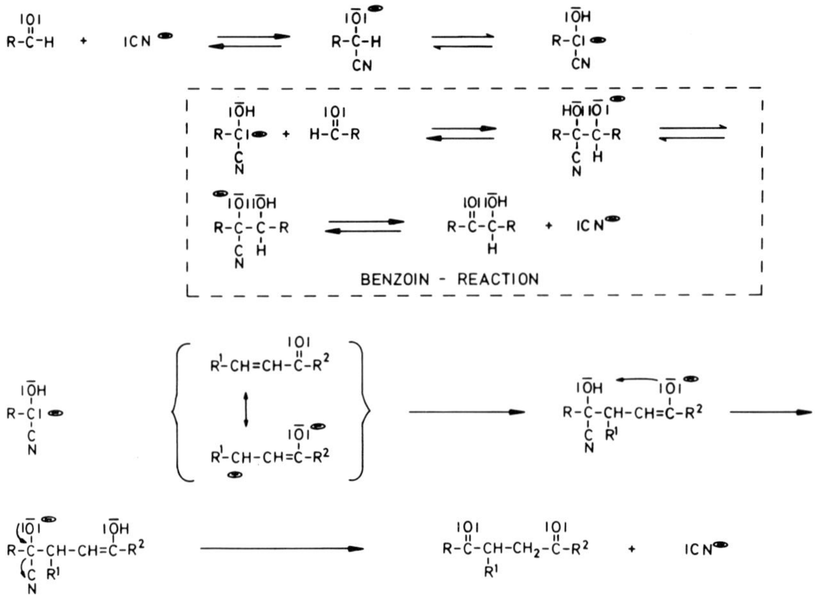

Fig. 1 The benzoin reaction, and the mechanism for the addition of aromatic and heterocyclic aldehydes to the activated double bond of α,β-unsaturated ketones (R^2=alkyl) or esters (R^2=O-alkyl).

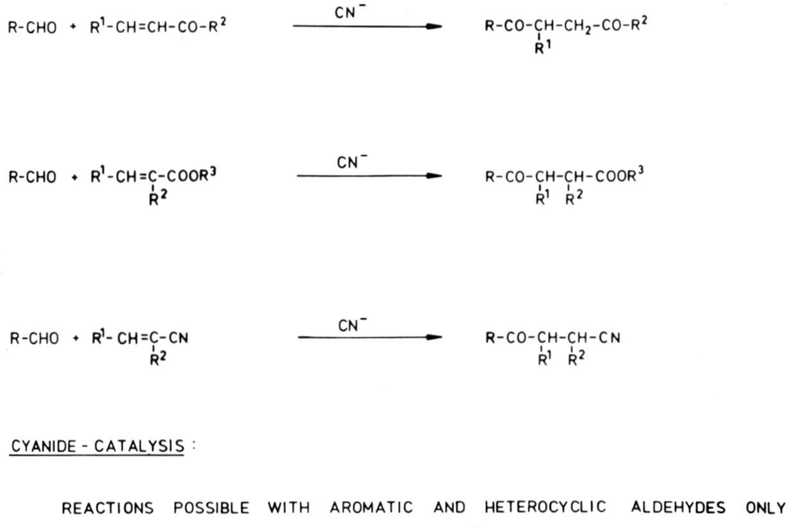

Fig. 2 The cyanide ion-catalyzed addition of aromatic and heterocyclic aldehydes to the activated double bond of α,β-unsaturated ketones, esters and nitriles to yield γ-diketones, 4-oxocarboxylic acid esters and 4-oxonitriles, respectively.

Catalysis with Thiazolium Salts

Biochemical studies have shown that thiamine (vitamin B_1) can convert aliphatic aldehydes into α-hydroxyketones (acyloins) in buffered aqueous solutions [7, 8]. The catalytic activity is associated with the thiazolium moiety of thiamine [8]. It is also known that thiazolium salts, which are unsubstituted at the 2-position, can generally assume this catalytic function [8].

In the presence of a base, quaternary thiazolium salts are transformed into a ylide structure. The ylide is then able to exert a catalytic effect resembling that of the cyanide ion in the benzoin reaction, and thus promote a similar addition reaction. Thiazolium salts, in the presence of base, have been successfully employed as catalysts for the addition of aliphatic, aromatic and heterocyclic aldehydes to α,β-unsaturated ketones, esters and nitriles [9]. The reaction yields γ-diketones, 4-oxocarboxylic acid esters and 4-oxonitriles, respectively (Fig. 3).

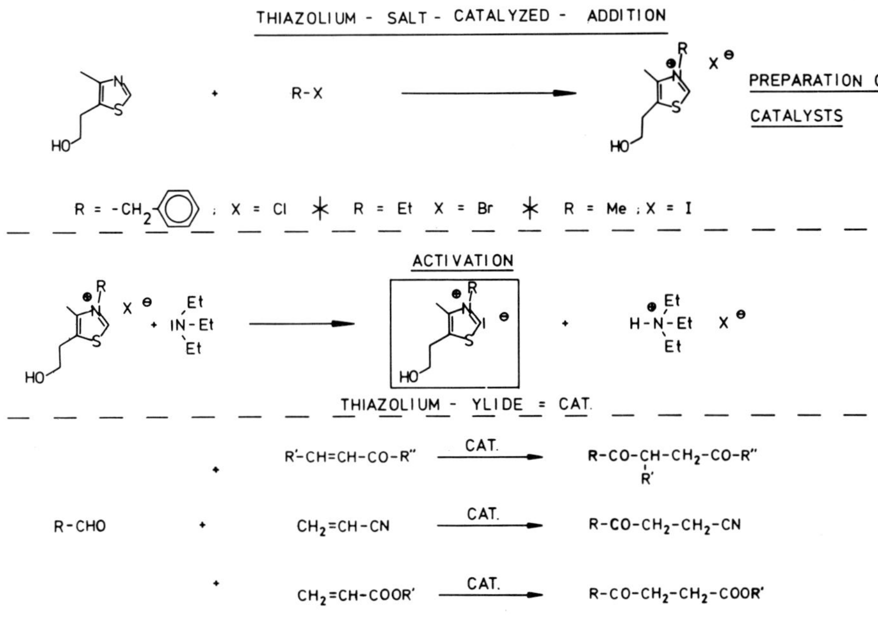

Fig. 3 Preparation and activation of the thiazolium ylide catalyst and its use in the addition of aldehydes to α,β-unsaturated ketones, nitriles and esters to yield γ-diketones, 4-oxonitriles and 4-oxocarboxylic acid esters, respectively.

The catalytic activity of many thiazole derivatives was investigated [10]. None of these quaternary salts, other than those of 5-(2-hydroxyethyl)-4-methyl-1,3-thiazole, showed any specific advantage, so only salts of this inexpensive and commercially available thiazole (obtained from thiamine production) were used. With aliphatic aldehydes, *N*-benzylated salts such as 3-benzyl-5-(2-hydroxyethyl)-4-methyl-1,3-thiazolium chloride proved most suitable [11], while aromatic aldehydes responded best to *N*-alkylated salts such as 3-ethyl-5-(2-hydroxyethyl)-4-methyl-1,3-thiazolium bromide [12] or 5-(2-hydroxyethyl)-3,4-dimethyl-1,3-thiazolium iodide [13]. Both kinds of catalysts were used for heterocyclic aldehydes.

The most suitable bases for the thiazolium salt-catalyzed reaction were triethylamine and sodium acetate. Both protic solvents, such as ethanol, and aprotic solvents, such as dimethylformamide or dioxane, could be used, while in some cases it was best to use no solvent at all. In general, reactions were carried out at 60–80 °C for periods of from 6 to 15 hr. The amount of catalyst was usually 10 mole percent and the amount of base 10 to 30 mole percent, when using 1 mole of aldehyde.

Applications of the Reaction

A great variety of aliphatic, heterocyclic and aromatic aldehydes can be employed in the thiazolium salt-catalyzed addition [9], and some examples are listed in Table 1. The reaction was also tested on a variety of unsaturated systems, and found to be rather widely applicable. In some cases, however, the reaction failed to proceed satisfactorily. Thus, for example, while acrylic esters and acrylonitrile readily undergo the addition, the presence of substituents on the double bond of these simple unsaturated esters and nitriles prevents any reaction. There are a few exceptions to this generalization, however, in that α-methylenbutyrolactone [20] readily adds aldehydes, as do alkylidene malonate and acetoacetate esters [9, 21], which are systems of higher activation.

When dealing with the monoactivated α,β-unsaturated ketones, the situation is not so strictly limited. Thus, monosubstituted systems react satisfactorily, but di- and tri-substituted systems fail to yield addition products. The most common unsaturated ketone used in these

Table 1 Aldehydes Employed in the Thiazolium Salt-Catalyzed Addition Reaction.

1. Straight-chain aliphatic aldehydes (from C_2 to C_{12} were tested).
2. Branched aliphatic aldehydes such as 2-methyl-propanal or 2-ethyl-butanal can be used successfully, but not pivalaldehyde (probably due to steric hindrance).
3. Conjugated unsaturated aldehydes, such as β,β-dimethylacrolein, citral and cinnamic aldehyde [14].
4. Aldehydes that contain isolated double bonds such as 10-undecenal, citronellal, 3-cyclohexen-1-carbaldehyde and norbornene aldehyde [14].
5. Araliphatic aldehydes such as hydrocinnamic aldehyde.
6. Aldehydes containing ether groups, including diverse ethers of glycolic aldehyde or β-alkoxy-propionic aldehyde [15, 16].
7. Aldehydes containing the phthalimido group such as phthalimidoacetaldehyde or phthalimido-propionaldehyde [17].
8. Aldehydes that contain additional functional groups such as keto-, ester- or nitrile groups (eg., levulinic aldehyde and succinic aldehyde ester) [18].
9. Heterocyclic aldehydes such as furan-2-carbaldehyde, thiophene-2-carbaldehyde and pyridine 2-, 3- or 4-carbaldehydes [19].
10. Aromatic aldehydes such as benzaldehyde, the tolylaldehydes, the anisaldehydes, *o*-, *m*- or *p*-chlorobenzaldehydes and naphthaldehydes.

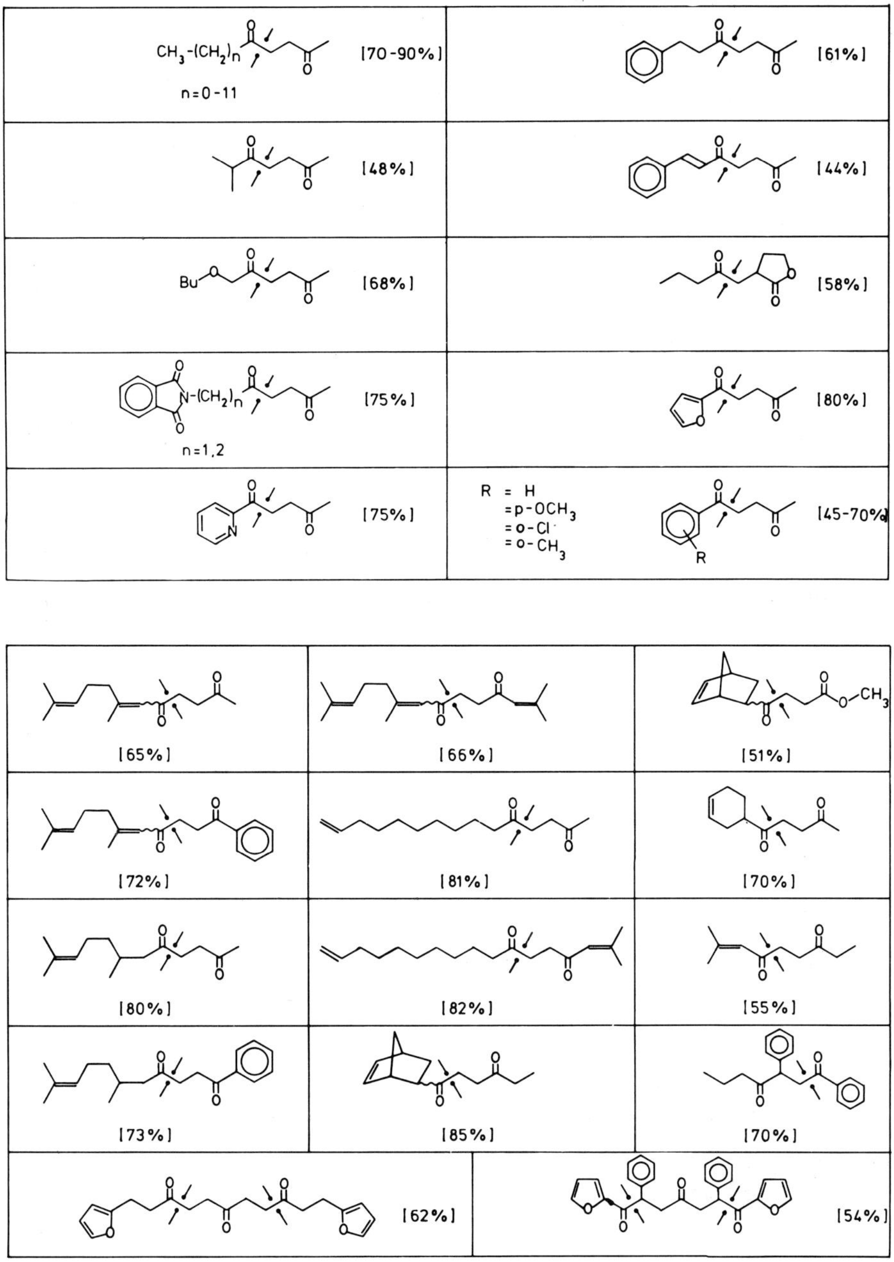

Fig. 4 Representative γ-oxo-compounds prepared by the thiazolium salt-catalyzed addition reaction. The generalized structure R–CO–CR′H-CH_2–CO–R″ indicates a compound synthesized from R–CHO + R′–CH=CH–CO–R″.

studies was butenone (methyl vinyl ketone), and this compound reacts successfully with all the classes of aldehydes previously mentioned. Furthermore, higher homologues of butenone have also been used (e.g., pent-1-en-3-one, hex-1-en-3-one, *t*-butyl vinyl ketone and phenyl vinyl ketone). The functional groups of the unsaturated ketones are, in principal, the same as those allowable in the chain or ring of an aldehyde (e.g., ether-, amido-, ester-, keto- or nitrile functions). Other unsaturated ketones successfully used were chalcones, ketones which bear one or two different aromatic or heterocyclic substituents, benzylidene acetone, and pent-3-en-2-one. Another source of unsaturated ketones used were Mannich bases and methyl ammonium salts [22]. Figure 4 illustrates a selection of representative γ-oxo-compounds prepared by the procedure described.

While the direct addition of acrolein would seem to be impossible, the reactive double bond of acrolein can be protected in a reversible manner using the Diels-Alder reaction with cyclopentadiene. The resulting norbornene aldehyde adds very well, and the addition product is cleavable by pyrolysis. α,β-unsaturated γ-diketones synthesized in this way may be used for further addition reactions, resulting in 1,4,7-triketone products [23]. Other methods for the production of tri- and polyketones have also been developed [13], including the addition of aldehydes to divinylketone or dibenzylidenacetone, leading to symmetrical triketones (Fig. 4). Symmetrical γ-diketones can be obtained by the reaction of aldehydes with α,β-unsaturated sulfones [24]. The α, β-unsaturated ketone is generated *in situ* from the aldehyde and the sulfone, for example divinylsulfone, and adds the aldehyde in the usual manner.

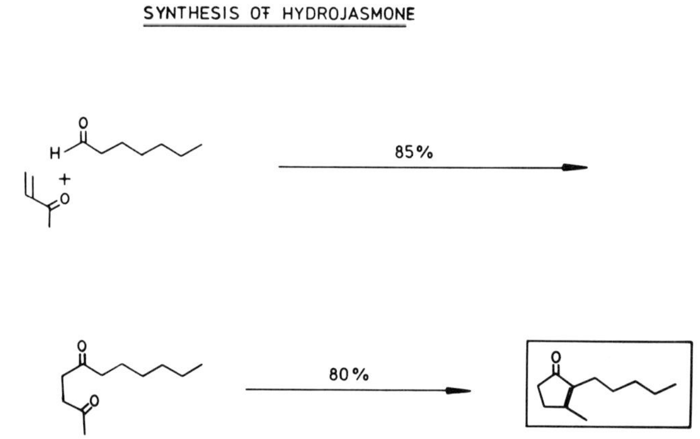

Fig. 5 Synthesis of hydrojasmone by the addition of heptanal to butenone followed by cyclization of the resulting 2,5-undecandione.

Synthesis of Jasmine Fragrances and Related Compounds

The γ-oxo-compounds, resulting from the addition reaction, are starting materials for the synthesis of legions of subsequent products including heterocyclic compounds such as furans, thiophenes, pyrroles, pyridazinones and γ-lactones, to name but a few simple monocyclic types.

The possibility of obtaining cyclopentenones by cyclizing aldol condensation [25] of γ-diketones is of particular interest. The synthesis of the group of jasmine odors has been the goal of innumerable efforts [26–29], and the reaction described above, which is catalytic in nature and uses simple and relatively inexpensive starting materials, should have unique utility in producing synthetic raw materials for these fragrances.

Initial efforts in this field were directed toward the synthesis of hydrojasmone [11] (Fig. 5). The two-step synthesis is very simple; first heptanal is added to butenone, and the resulting 2,5-undecandione is then cyclized with alkali by the method of Hunsdiecker [25]. While this reaction was performed on a hundred gram level in the laboratory, the scale-up for large quantity technical preparation seems to be economically feasible. In a subsequent attempt to synthesize *cis*-jasmone [11], the problem was not the addition, but rather the synthesis of the required aldehyde (Fig. 6). In laboratory trials, possibility A (using *cis*-3-hexen-1-ol) was successfully tried. Possibility B, which would appear to be more economical, remains to be tested.

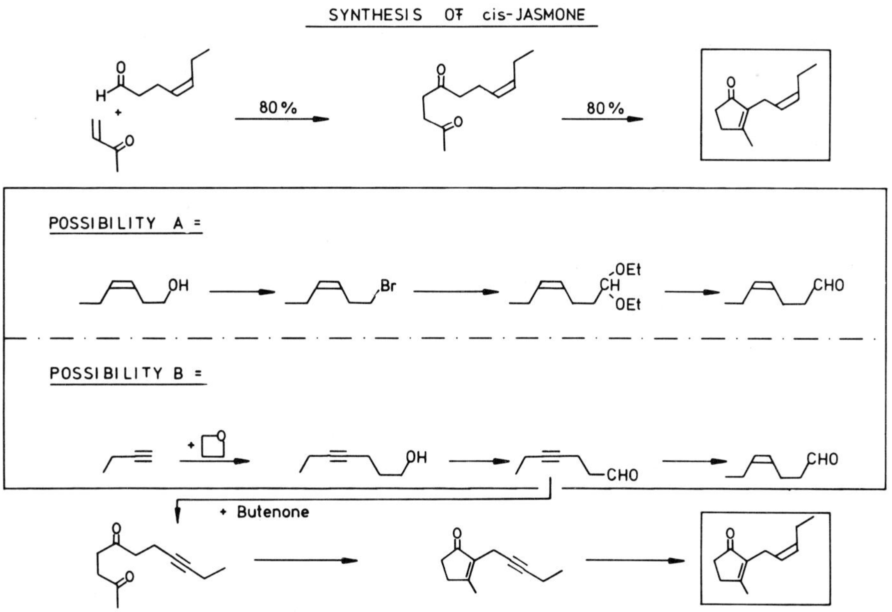

Fig. 6 Possible alternatives for the synthesis of *cis*-jasmone.

The most important jasmine odiferous substance, methyl dihydrojasmonate (otherwise known as hedione), has also been synthesized (Fig. 7). The first step of the synthesis is the addition of heptanal to ethyl acrylate or acrylonitrile. Saponification gives 4-oxodecanoic acid, which is the key intermediate for the synthesis. The latter is usually synthesized via radical addition of heptanal to diethyl maleate and subsequent saponification and decarboxylation [30]. The initial problem was the addition of the aldehyde to an ester or nitrile, as previously mentioned. By variation of the *N*-alkyl substituent of the catalyst, the yields obtained for this

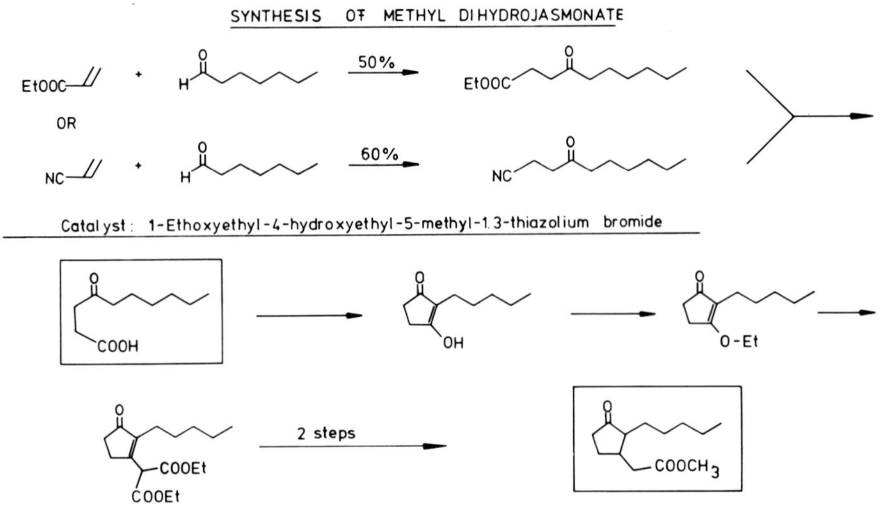

Fig. 7 Synthesis of methyl dihydrojasmonate by the addition of heptanal to ethyl acrylate or acrylonitrile to form 4-oxodecanoic acid, followed by cyclization and subsequent steps.

crucial step were optimized. The best catalyst so far is 3-(β-ethoxyethyl)-5-hydroxyethyl-4-methyl-1,3-thiazolium bromide developed by Stetter and Nienhaus [20]. The resulting γ-keto ester (Fig. 7) and its homologues can be obtained in yields as high as 50 %, and the corresponding nitriles in yields as high as 60%. The use of bulk chemicals makes this synthesis of 4-oxodecanoic acid preferable to others. By reduction of the oxo-group of such γ-keto acids, γ-alkylbutyrolactones become available (Fig. 8). These compounds are not only odoriferous substances themselves [31], but are also a source of alkyl cyclopentenones [32].

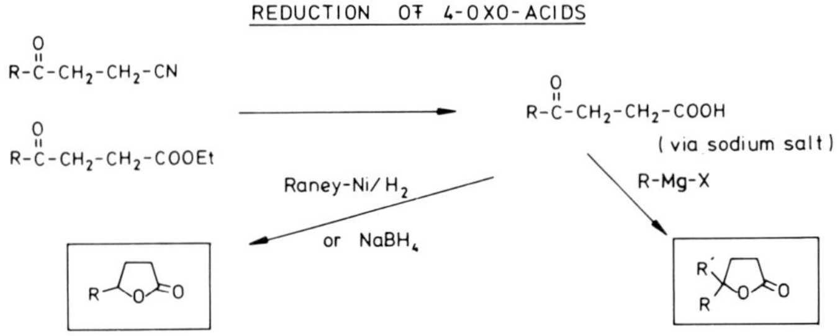

Fig. 8 Reduction of 4-oxo-acids to afford γ-alkylbutyrolactones.

In addition to methyl dihydrojasmonate, an isomeric methyl isodihydrojasmonate can be constructed by changing both substituents [18] (Fig. 9). Thus, reaction of succinic aldehyde ester with pentyl vinyl ketone, and subsequent cyclization of methyl 4,7-dioxododecanoate followed by hydrogenation affords methyl isodihydrojasmonate. Both the cyclopentenone and the cyclopentanone (Fig. 9) exhibit an odor similar to hedione.

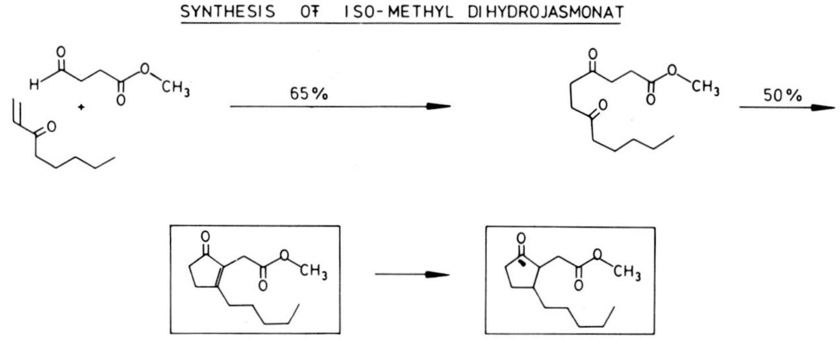

Fig. 9 Scheme for the synthesis of methyl isodihydrojasmonate via the addition of succinic aldehyde methyl ester to pentyl vinyl ketone, followed by cyclization of the resulting methyl 4,7-dioxododecanoate and subsequent hydrogenation.

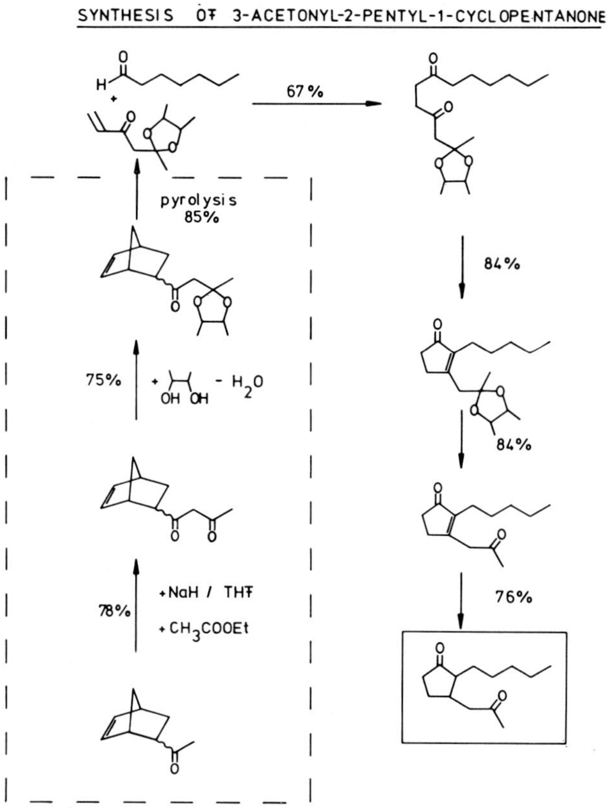

Fig. 10 Scheme for the synthesis of 3-acetonyl-2-pentyl-1-cyclopentanone (details of the method can be found elsewhere [33]).

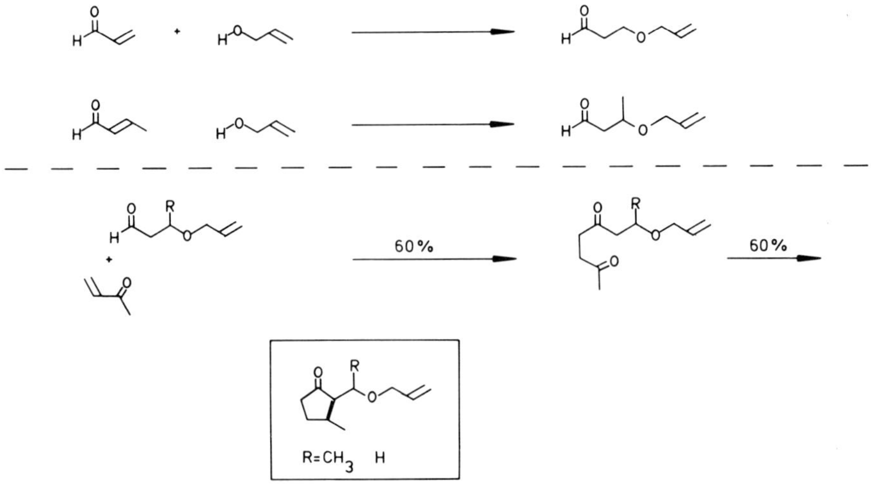

Fig. 11 Scheme for the synthesis of two new cyclopentenones (R=CH_3 or H). The required aldehydes, prepared by Michael addition of allyl alcohol to acrolein or crotonaldehyde respectively, are reacted with butenone and the products cyclized.

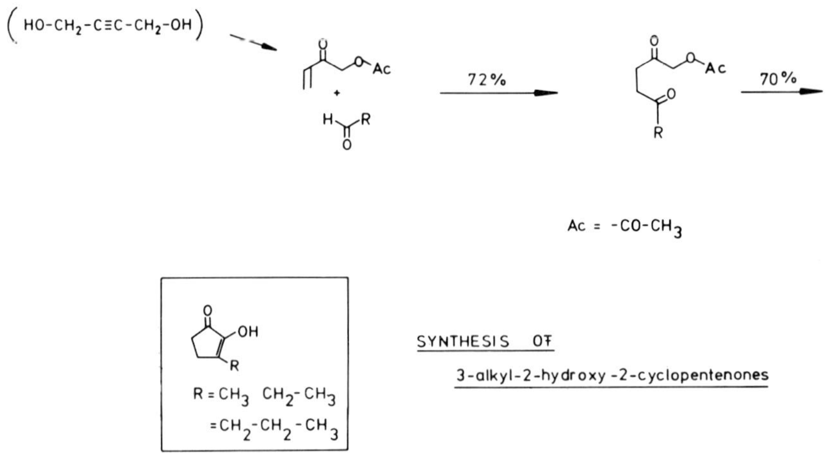

Fig. 12 Scheme for the synthesis of 3-alkyl-2-hydroxy-2-cyclopentenones by addition of the appropriate aldehyde (R=methyl, ethyl or propyl) to acetoxymethyl vinyl ketone (available from butynediol) to afford the corresponding γ-diketone which is subsequently cyclized.

The procedure described herein was also utilized to synthesize 3-acetonyl-2-pentyl-1-cyclopentanone [33] as outlined in Figure 10. This cyclopentanone derivative has fragrance qualities similar to methyl dihydrojasmonate, having both a strong magnolia odor and good fixative properties [34]. While the present synthesis [33] is of purely academic interest, the fragrance properties of this compound would justify attempts to develop a simpler synthesis, perhaps via the addition of ethyl acetoacetate to pentylcyclopentenone. The homologous hexyl compound (3-acetonyl-2-hexyl-1-cyclopentanone), having similar characteristics, has also been synthesized.

The structures of two new odoriferous substances, closely related to the jasmine scents but bearing an additional striking nut odor, are shown in Figure 11 (R = CH_3 or H). The compound with the methyl substituent has the more distinctive odor. The required aldehydes are prepared by Michael addition of allyl alcohol to acrolein and crotonaldehyde, respectively. Addition reactions with butenone, followed by alkaline cyclization of the products, leads to the new cyclopentenones [35] (Fig. 11).

Another class of substances which are important natural flavor constituents, and which has become available through this synthetic procedure, is the 3-alkyl-2-cyclopenten-2-ol-1-ones [36]. Addition of aldehydes to acetoxymethyl vinyl ketone, which is readily available from butynediol, leads to γ-diketones in 70% yield, and subsequent cyclization leads to the flavor compounds [37] (Fig. 12).

Acknowledgement

We would sincerely like to thank all our colleagues who worked with us in the field of aldehyde additions during the past few years.

References

[1] Ide, W. S., and J. S. Buck, Organic Reactions **4**, 269 (1949).
[2] Kuebrich, J. P., R. L. Schowen, M.-S. Wang and M. E. Lupes J. Am. Chem. Soc. **93**, 1214 (1971).
[3] Stetter, H., and M. Schreckenberg, Chem. Ber. **107**, 2453 (1974).
[4] Stetter, H., M. Schreckenberg and K. Wiemann, Chem. Ber. **109**, 541 (1976).
[5] Stetter, H., and M. Schreckenberg, Chem. Ber. **107**, 210 (1974).
[6] Kuhlmann, H., Dissertation TH Aachen 1976.
[7] Schilling, C. L., and J. E. Mulvaney, Macromolecules **1**, 452 (1968).
[8] Breslow, R., J. Am. Chem. Soc. **80**, 3719 (1958).
[9] Stetter, H., Angew. Chem. **88**, 695 (1976); Angew. Chem. Int. Ed. Engl. **15**, 639 (1976).
[10] Hilboll, H., Diplomarbeit TH Aachen 1975.
[11] Stetter, H., and H. Kuhlmann, Synthesis **1975**, 379.
[12] Stetter, H., and H. Kuhlmann, Chem. Ber. **109**, 2890 (1976).
[13] Stetter, H., W. Basse, H. Kuhlmann, A. Landscheidt and W. Schlenker, Chem. Ber. **110**, 1007 (1977).
[14] Stetter, H., G. Hilboll and H. Kuhlmann, Chem. Ber. **112**, 84 (1979).
[15] Stetter, H., and J. Nienhaus, Chem. Ber. **111**, 2825 (1978).
[16] Stetter, H., K. Mohrmann and W. Schlenker, Chem. Ber. in press.
[17] Stetter, H., and P. Lappe, Chem. Ber. in press.
[18] Stetter, H., W. Basse and K. Wiemann, Chem. Ber. **111**, 431 (1978).
[19] Stetter, H., and J. Krasselt, J. Heterocyclic Chem. **14**, 573 (1977).
[20] Stetter, H., and J. Nienhaus, Chem. Ber. **113**, 690 (1980).
[21] Jonas, F., Dissertation TH Aachen 1979.
[22] Stetter, H., P. H. Schmitz and M. Schreckenberg, Chem. Ber. **110**, 1971 (1977).

[23] Stetter, H., and A. Landscheidt, Chem. Ber. **112**, 1410, 2419 (1979).
[24] Stetter, H., and H.-J. Bender, Angew. Chem. **90**, 130 (1978).
[25] Hunsdiecker, H., Ber. d. dtsch. chem. Ges. **75**, 447 (1942).
[26] Van der Gen, A., Parf. Cosm. Sav. (France) **2**, 356 (1972).
[27] Joulain, D., Parf. Cosm. Aromes **4**, 33 (1975).
[28] Joulain, D., Parf. Cosm. Aromes **12**, 53 (1976).
[29] Dubs, P., and R. Stüssi, Helv. Chim. Acta **61**, 990 (1978).
[30] Patrick, T. M. Jr., and F. B. Erickson, Org. Syntheses Coll. Vol. **4**, 430 (1963).
[31] v. Braun, J., Ber. d. dtsch. chem. Ges. **70**, 1250 (1937).
[32] Plattner, Pl. A., and A. St.-Pfau, Helv. Chim. Acta **20**, 1474 (1937).
[33] Nienhaus, J., Dissertation TH Aachen 1979.
[34] Aebi, H., E. Baumgartner, H. P. Fiedler and G. Ohloff, Kosmetika, Riechstoffe und Lebensmittelzusatzstoffe, G. Thieme Verlag (Stuttgart) 1978, p. 60.
[35] Stetter, H., and G. Hilboll, Synthesis **1979**, 187.
[36] Barco, A., S. Benetti, G. P. Pollini and R. Taddia, Synthesis **1975**, 104.
[37] Schlenker, W., Dissertation TH Aachen 1979.

9. PROGRESS IN THE SYNTHESIS OF THE SANDALWOOD FRAGRANCE, 3-*trans*-ISOCAMPHYLCYCLOHEXANOL

G. K. Lange* and K. A. Bauer*

Summary A brief review describes both the historical development of the industrially produced sandalwood fragrance mixture and the investigation which led to the discovery of the active principle 3-*trans*-IC-cyclohexanol (*4*, IC = isocamphyl). The present manufacturing process can be significantly improved by carrying out an extractive separation of *p*-IC-guaiacol (*1*) from the so-called bornylguaiacol that is formed through the alkylation of guaiacol with camphene.

The former compound (*1*) is esterified to (*2*, X = ester) and, after removal of the phenolic hydroxyl group and cleavage of the methylether (*3*, R = CH_3, Z = H), *m*-isocamphylphenol (*3*, R = Z = H) is obtained in high yield. Subsequent hydrogenation gives mixtures with a high content of 3-*trans*-isocamphylcyclohexanol (*4*).

Scheme 1

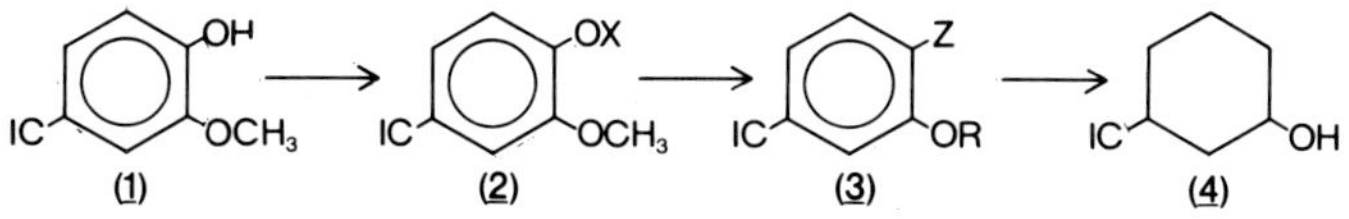

Alternatively, compound (*1*) can be ethylated to give (*2*, X = C_2H_5) which then undergoes selective demethylation to yield *m*-IC-guaethol (*3*, R = H, Z = OC_2H_5). Hydrogenation of the latter with concomitant elimination of the ethoxy group gives the sandal compound (*4*) directly.

Introduction

The synthesis of the industrial sandalwood fragrance, which was first discovered some 40 years ago, competes effectively with the naturally occurring East Indian sandalwood oil, and both of these products play very important roles as perfumery materials. As the name implies,

* G. K. Lange and K. A. Bauer, Haarmann & Reimer GmbH, 3450 Holzminden, West Germany.

sandalwood oil is obtained from the wood of the sandalwood tree, and it was well-known even in antiquity [1]. The very pleasant, persistant woody fragrance was held in great esteem. Presumably, the wood was used initially for religious purposes and then later the oil was employed as a cosmetic. Very pleasant smelling oils, which were used for anointing, were known in all the ancient civilizations especially those in India, Egypt, Greece and Rome. A reference to the value of sandalwood in ancient times can be found in the First Book of Kings in the Old Testament [2]. It recalls how King Solomon received a large quantity of sandalwood which he used in building the temple and his royal palace. The wood came from the mysterious land of Ophir which lay across the Red Sea. This happened some 3000 years ago, and sandalwood has remained irreplaceable ever since.

Structure of Natural and Synthetic Sandalwood Fragrance Compounds

The structural elucidation of the main components of sandalwood oil was achieved at the beginning of this century. The two sesquiterpene alcohols, α- and β-santalol (Fig. 1) make up more than 90 % of the oil [3]. These two alcohols are related to each other in the same way as tricyclene is to camphene. Furthermore, the side chain is the same in both cases with the double bond in the *cis*-configuration. The total synthesis of these alcohols proved to be very difficult, and the introduction of the *cis*-hydroxymethyl grouping was especially troublesome. This problem was finally solved by E. J. Corey who, by means of the Wittig reaction, synthesized the unsaturated alcohol with the necessary *cis*-configuration [4]. The odor properties of the industrial sandalwood fragrance are dependent, to a certain extent, upon the configuration of the alcohol grouping [5], as is also the case with the natural santalols [6].

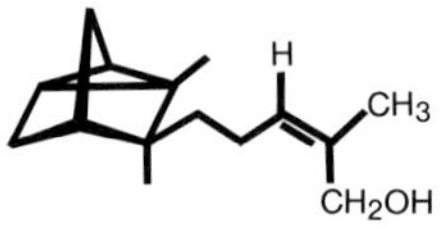

α-Santalol

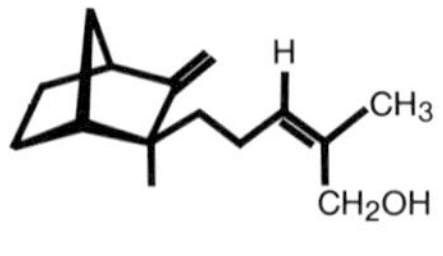

β-Santalol

Fig. 1 Structures of α-santalol and β-santalol.

East Indian sandalwood oil has been one of the most precious perfumery materials from antiquity to modern times, and its popularity shows no sign of waning. It is, however, not available in unlimited quantities, and so one can imagine the delight of the organic chemist who discovered that he had synthesized a material that had a strong sandalwood odor. This person was Dr. Albert Weissenborn who, while working for the IG Farben in Wolfen, caused camphene to react with guaiacol in the presence of boron trifluoride. The odorless reaction product was then hydrogenated at high temperatures using Raney nickel, and it was the product of this reaction that possessed the strong sandalwood odor. These findings were submitted for patent protection; the first patent being filed in 1942 and covering the hydrogenation of terpene phenols in general. The second patent, filed in 1943, claimed protection for the hydrogenation of terpene phenols which not only had a free hydroxyl group but also one in the form of an ether. However, due to the confusion and turmoil during and after the war, these patents were first published in 1952 [7]. Weissenborn probably knew very little about the actual composition of his product. Terpenyl phenols were known at that time, but the analytical techniques available were not able to provide information about the identity of the terpene residue or about the attachment to the phenol (or the derived cyclohexane ring).

Towards the end of the 1950's, a number of investigators examined the composition of this very complex reaction mixture. Both the Russian chemist Kheifits and co-workers [8–11] and various American groups [12–15] were involved in these studies. Thus, it became apparent from a U.S. patent granted to Givaudan [16], that the odorous components were not hexahydroguaiacols, but rather substances which arose by hydrogenolysis of the ether group of the guaiacol moiety. However, the substance actually responsible for the fragrance of the reaction mixture was not identified and it was not known whether camphene had undergone rearrangement before reaction with guaiacol. It was later shown that the main products of the reaction were isocamphyl- and isobornylguaiacol, and that isofenchyl skeletons were also present [17] (Fig. 2). These identifications were made by isolation of the individual compounds followed by oxidative degradation using potassium permanganate to form ketones of known structure [18] (Fig. 3). Nowadays, the structural assignments can be achieved relatively easily by means of NMR. The carbon skeletons differ quite distinctly, as can be seen in the 360 MHz spectra (Fig. 4).

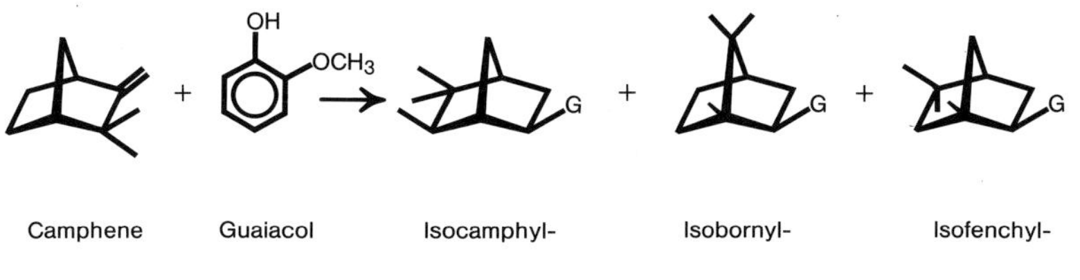

Fig. 2 Products formed on condensation of camphene with guaiacol. G represents the guaiacol moiety.

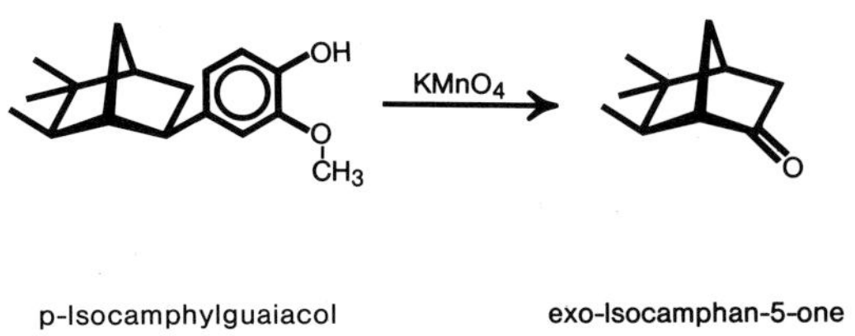

Fig. 3 Oxidation of *p*-isocamphylguaiacol to *exo*-isocamphanone.

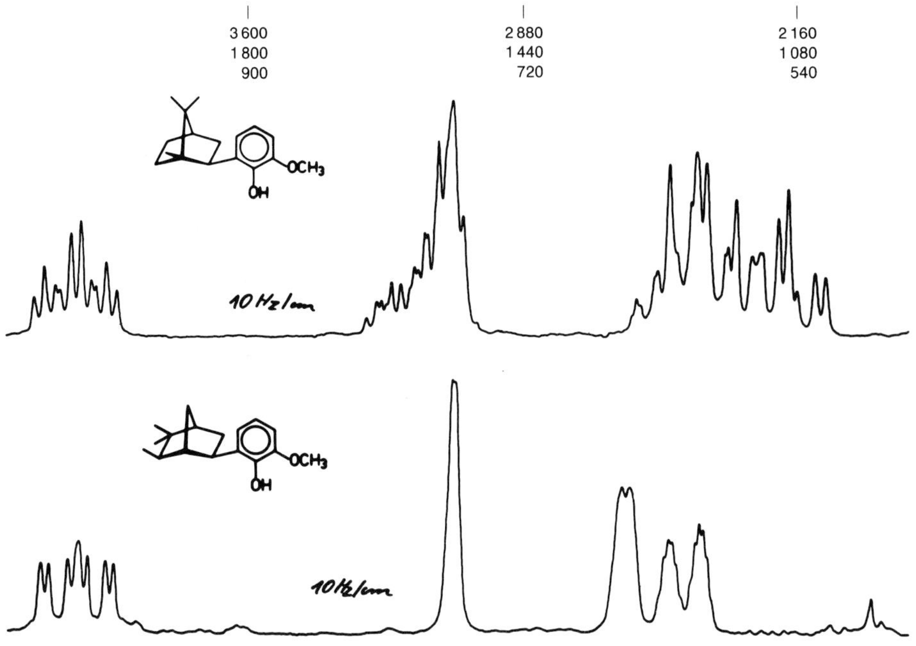

Fig. 4 360 MHz NMR spectra of Isobornyl guaiacol (upper) and isocamphyl guaiacol (lower).

Theoretically, these three terpene residues (isocamphyl-, isobornyl-, isofenchyl-) can occupy the four available positions of the guaiacol molecule (Fig. 5). As expected, the hydroxyl group exerts a stronger inductive effect upon the incoming groups than the methoxyl substituent. Consequently, one finds the terpene residue preferentially in the *ortho* and *para* positions relative to the hydroxyl group. However, as indicated below, the isomers of interest happen to be those which are formed as a result of the inductive effect of the methoxyl group!

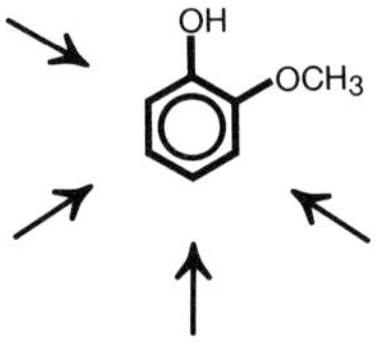

Fig. 5 Alkylation of guaiacol can theoretically occur at four positions.

Gas chromatographic analysis of the reaction products obtained by alkylation of guaiacol with camphene in the presence of boron trifluoride reveals a rather complex mixture (Fig. 6). The main components are *o*-isobornylguaiacol, *o*-isocamphylguaiacol and *p*- and *m*-isocamphylguaiacol. Not surprisingly, an even more complex mixture is obtained upon hydrogenation of the crude reaction products (Fig. 7).

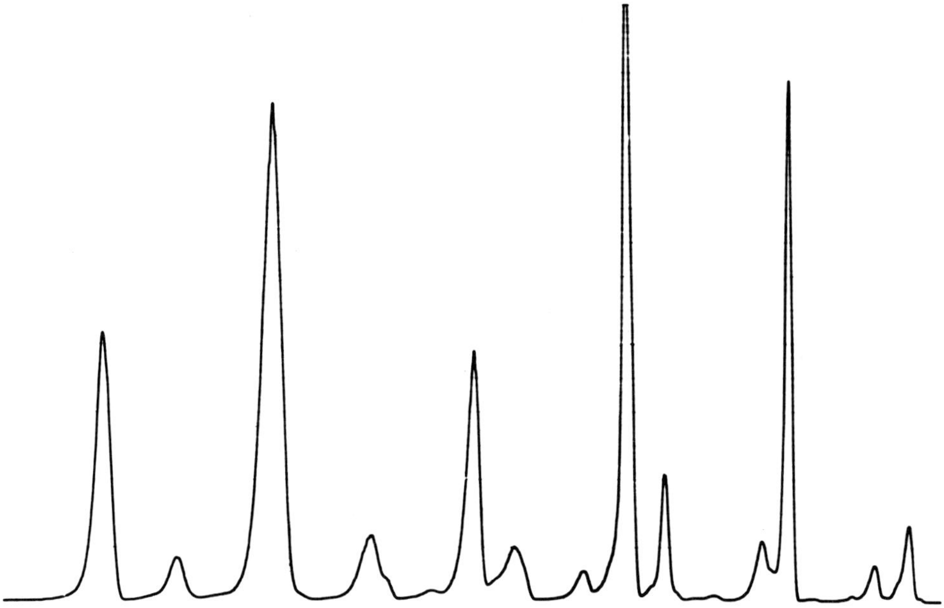

Fig. 6 Gas chromatographic analysis of the reaction products obtained by alkylation of guaiacol with camphene in the presence of boron trifluoride.

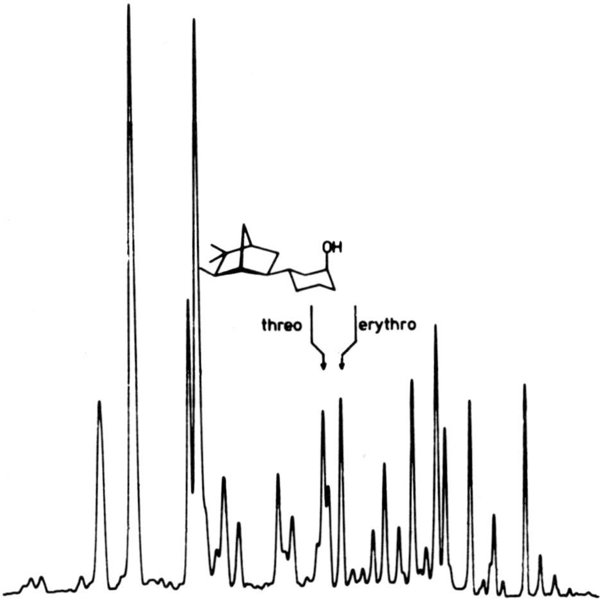

Fig. 7 Gas chromatographic analysis of the mixture obtained upon hydrogenation of the reaction products described in Fig. 6.

It is entirely due to the work of Edouard Demole that we now know which of these numerous reaction products exhibit the sandalwood fragrance properties. As the starting material in his investigations, Demole [5] used terpenyl phenols which he obtained by reacting phenol with camphene in the presence of boron trifluoride. The individual components of the reaction mixture were isolated in a pure state and subsequently hydrogenated. Demole found that four different terpenylcyclohexanols were formed when pure *o*-isocamphylphenol (*5*) was hydrogenated [19] (Fig. 8). Not only can the hydroxyl group be in either the axial or equatorial

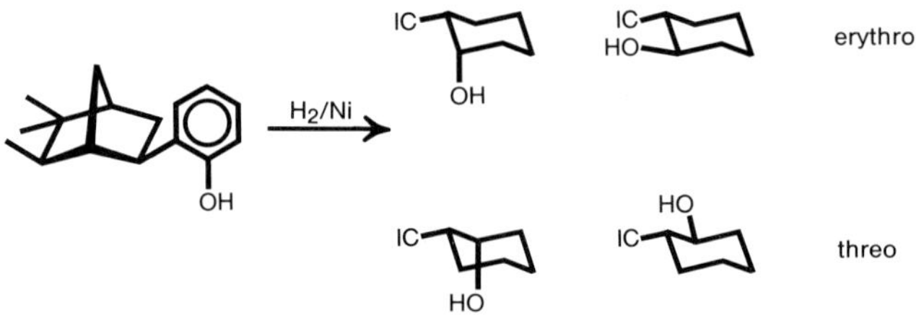

Fig. 8 Hydrogenation of *o*-isocamphylphenol [*5*] yields four isomeric isocamphylcyclohexanols (IC=isocamphyl).

position, but also on either the same side (*erythro*) or opposite side (*threo*) relative to the *exo*-methyl group of the terpene residue. Fortunately, the terpene residue is always in the equatorial position, otherwise there would be eight different compounds instead of four. The four compounds formed were, of course, racemic and when they were isolated in a pure state, they all proved to be odorless [20]. Eventually, Demole discovered that the 3-*trans*-isocamphylcyclohexanols (Fig. 9) were responsible for the sandalwood fragrance [5]. He further

Fig. 9 The *erythro* and *threo* isomers of 3-*trans*-isocamphylcyclohexanol have similar odors.

showed that the isocamphyl group and an axial hydroxyl group in the 3 position were necessary for the odor response, and that the *erythro* and *threo* isomers were equally effective. All the other cyclohexanols substituted with the isobornyl-, isocamphyl-, and isofenchyl-skeletons were more or less odorless.

Comparison of the structures of the synthetic fragrance, 3-*trans*-isocamphylcyclohexanol, and the naturally occurring sandalwood fragrance, α-santalol, reveals few genuinely common features (Fig. 10). Thus, while a similarity between the geometry of the *cis*-chain of α-santalol and the cyclohexanol ring of the synthetic compound is apparent, this does not explain why only the isocamphyl-substituted structure is active.

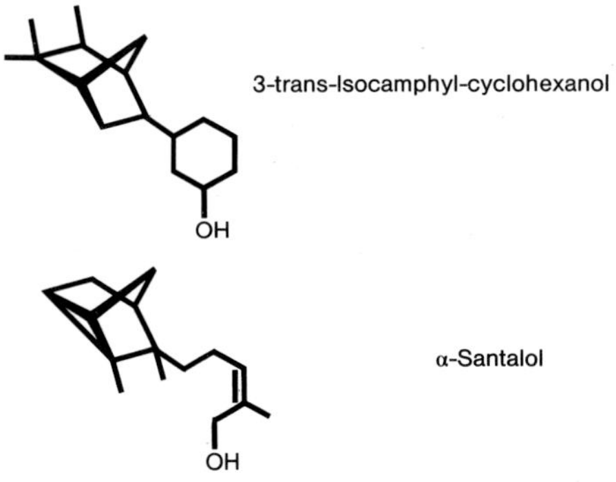

Fig. 10 Structures of the synthetic fragrance, 3-*trans*-isocamphylcyclohexanol, and the naturally occurring sandalwood fragrance, α-santalol.

Synthesis of the Sandalwood Fragrance

Analysis of the product mixture obtained from the industrial synthesis reveals approximately 8 % of the required isomer which possesses the sandalwood fragrance. The best method for determining the amount of this isomer present is gas chromatography of the corresponding acetates. The chromatogram illustrated (Fig. 7) was obtained using a 40 meter glass capillary column coated with Carbowax 20 M, and under these conditions a good separation of the *threo-* and *erythro-*components can be achieved. Additionally, the isocamphyl derivatives can be recognized by means of their doublet. The 8 % yield of the odoriferous component in the industrial sandalwood mixture is unsatisfactory, and it is necessary to find alternative, more efficient ways of synthesizing isocamphylcyclohexanols in which the hydroxy group is in position three. Furthermore, the stereochemistry of the C-3 hydroxyl group is important as only the axial epimer is active (the equatorial epimer is odorless).

Demole described a method for synthesizing 3-isocamphylcyclohexanol (*4*) using terpenylphenols as starting material [19] (Fig. 11). A reaction mixture containing approximately

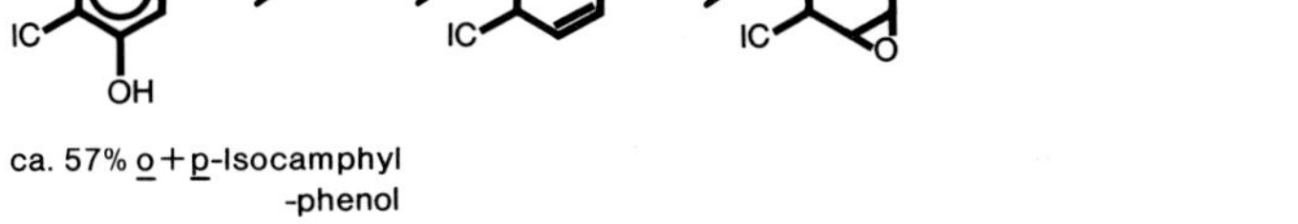

Fig. 11 Procedure for the synthesis of 3-isocamphylcyclohexanol (*4*) via isocamphylphenols (after Demole [19].

57% *ortho-* and *para*-isocamphylphenol was obtained from the boron trifluoride catalyzed alkylation reaction between phenol and camphene. This reaction mixture was then hydrogenated, acetylated and pyrolyzed. The cyclohexenes which were formed were converted into epoxides and the latter reduced to alcohols. Using this reaction scheme (Fig. 11) a certain amount of 3-isocamphylcyclohexanol is obtained from *ortho-* and *para*-isocamphylphenol. A difficulty, however, lies in the reduction of the epoxides, which are present as *cis/trans*-mixtures. The reaction conditions have to be chosen such that the epoxide ring opens to form an axial hydroxyl group at position three. Lithium aluminium hydride gives very good results, but the reagent is expensive and not easy to handle.

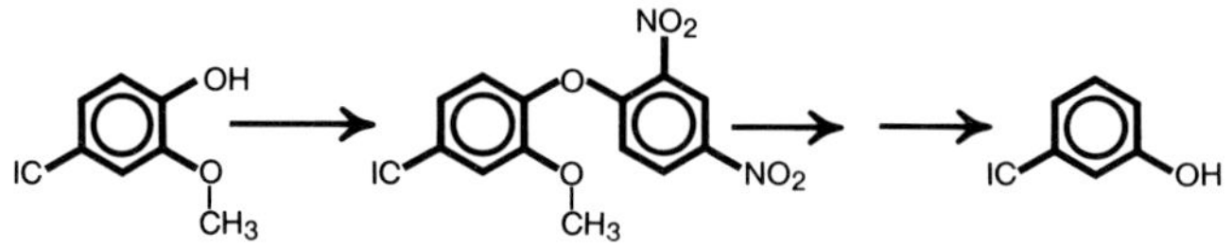

1 6

Fig. 12 Procedure for the conversion of *p*-isocamphylguaiacol (*1*) to *m*-isocamphylphenol (*6*) (after Kheifits [21]).

An alternate procedure, published by Kheifits [21], uses the reaction mixture obtained from guaiacol and camphene. (The reaction product has been incorrectly named "bornyl guaiacol".) *p*-Isocamphylguaiacol (*1*) can be extracted from the mixture and subsequently converted into *m*-isocamphylphenol (*6*) via a number of reaction steps (Fig. 12). The reaction scheme involves eliminating the hydroxyl group in the *para*-position relative to the terpene residue, cleaving the *meta*-isocamphylanisole to form *meta*-isocamphylphenol (*6*), and then hydrogenation of the latter. This reaction scheme not only requires dinitrofluorobenzene, which is expensive, but it also gives a poor overall yield due to the large number of reaction steps involved. Consequently, this process is of no industrial significance. However, if one was able to transform the *p*-isocamphylguaiacol into the *m*-isocamphylphenol by using a relatively simple series of reactions, then the process could be industrially significant. The following section reviews two viable procedures which allow the oxygen function in the *para*-position (relative to the isocamphyl residue) to be eliminated while that in the *meta*-position is retained (as in Fig. 13).

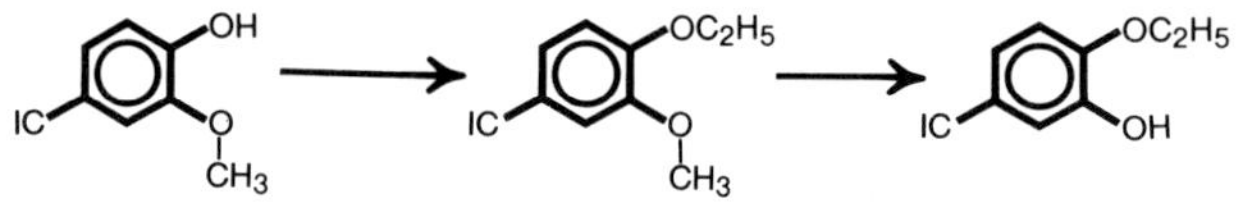

1 7 8

Fig. 13 Conversion of *p*-isocamphylguaiacol (*1*) to *m*-isocamphylguaethol (*8*) via the mixed ether (*7*).

The first procedure (Fig. 13) consists of using diethyl sulfate to convert *p*-isocamphylguaiacol (*1*) into a mixed ether (*7*) which undergoes nucleophilic cleavage to form *m*-isocamphylguaethol (*8*) [22]. The cleavage can be carried out with alcoholates or Grignard reagents, the choice of reagent determining the selectivity of the cleavage reaction which can provide about 95 % of the desired product (Fig. 14). In the presence of a suitable catalyst such as Raney nickel, the *m*-isocamphylguaethol (*8*) undergoes catalytic hydrogenation with concomitant loss of the ethoxy group, to afford the sandalwood fragrance in high yield. The fragrance intensity of the reaction mixture is directly proportional to the amount of 3-*trans*-isocamphylcyclohexanol present, which consequently determines its value. The choice of catalyst is of great importance here, as the catalyst determines the ratio of axial to equatorial alcohol formed.

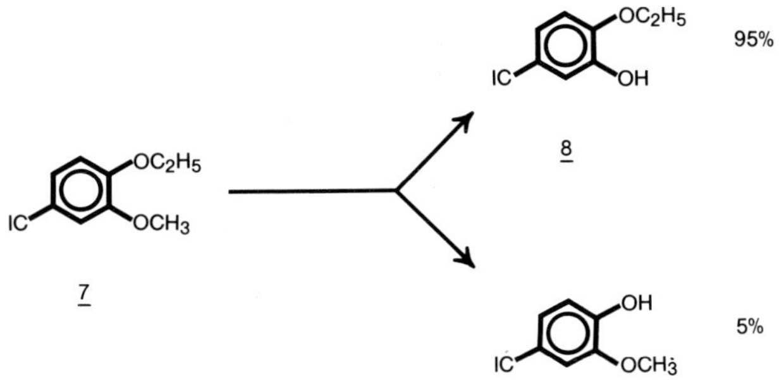

Fig. 14 Cleavage of the diether (*7*) to preferentially yield *m*-isocamphylguaethol (*8*).

The procedure described above involves the elimination of the *para*-oxygen function at the cyclohexanol step. The other possibility would be elimination of this oxygen function at the phenol step. During the last few years, a number of procedures have been published which enable suitably derivatized phenolic hydroxyl groups to undergo such hydrogenolytic elimination. The basic principle of this method entails the formation of derivatives which upon cleavage give rise to resonance stabilized molecules, thus favoring the cleavage reaction. Suitable derivatives are the esters of methane sulfonic acid [23] or cyanuric acid [24] and substituted isoureas [25] (Fig. 15). The hydrogenation reaction is carried out at room temperature under normal pressure, and uses palladium as catalyst. Any acids that may be formed can be trapped by adding an amine. In the present case, the product of the reaction is *m*-isocamphylanisole, which can be cleaved to the corresponding phenol using known procedures.

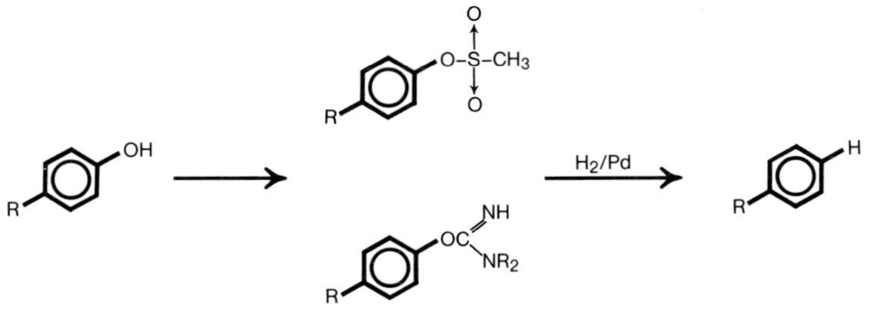

Fig. 15 Hydrogenolytic cleavage of the suitably derivatized phenolic hydroxyl group.

An especially suitable alternative to catalytic cleavage is the reduction of substituted phosphoric acid esters with sodium in liquid ammonia [26] (Fig. 16). The superior assets of this procedure are that the phosphates can be obtained in almost quantitative yields and that the cleavage reaction also gives high yields [22]. The product, *m*-isocamphylanisole, can then be converted in good yield into *m*-isocamphylphenol (*6*) using known procedures and the phenol can be hydrogenated specifically to the axial alcohol. Thus, the processes which provide *m*-isocamphylphenol show the distinct advantage that, given the correct choice of catalyst for the hydrogenation reaction, high yields of the required sandalwood fragrance can be obtained, and this automatically determines the fragrance intensity and value of the product.

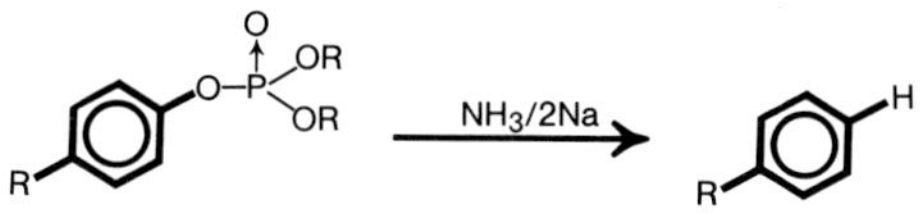

Fig. 16 An alternate means of removing the phenolic hydroxyl group by reduction of the substituted phosphoric acid ester.

It is obvious that pure axial alcohol cannot be obtained by hydrogenation of the phenol. However, the mixture of alcohols thus formed can be oxidized to the corresponding ketone, and the latter reduced specifically to the crystalline axial alcohol. Normal reductions using lithium aluminium hydride or sodium borohydride give mixtures of axial and equatorial alcohols in which the equatorial epimer predominates. Recently, however, reducing agents have been developed which form the axial alcohol specifically. These include the Henbest catalysts [27] and the bulky lithium alkyl borohydrides [28]. The Henbest catalyst is an iridium-phosphoric acid complex which is formed *in situ*, and its use for the reduction of 3-*t*-butylcyclohexanone has been described. In the present case, however, the speed of reduction of 3-isocamphylcyclohexanone (*9*) was slow and not worth following up. On the other hand, the bulky alkyl borohydrides, developed by H. C. Brown, were successful in reducing the ketone (*9*) to nearly pure axial alcohol (Fig. 17). This alcohol can be obtained in absolute purity by crystallization.

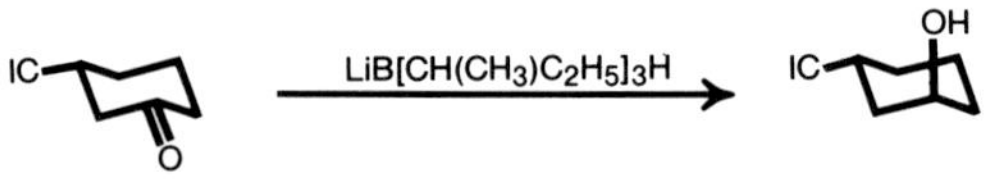

9

4

Fig. 17 Stereospecific reduction of 3-isocamphylcyclohexanone (*9*) to 3-*trans*-isocamphylcyclohexanol (*4*) with lithium alkyl borohydride.

References

[1] Gildemeister, E., and F. Hoffmann, Die ätherischen Öle, Vol. IV, p. 556, Akademie-Verlag, Berlin, 1956.
[2] Holy Bible, First Book Of Kings, Chapter 10, Verse 12.
[3] Gildemeister, E., and F. Hoffmann, Die ätherischen Öle, Vol. IIIb, p. 329, 4th Ed. (W. Treibs, ed.) Akademie-Verlag, Berlin, 1962.
[4] Corey, E. J., and H. Yamamoto, J. Am. Chem. Soc. **92**, 226 (1970).
[5] Demole, E., Helv. Chim. Acta **47**, 319, 1766 (1964).
[6] Lewis, R. G., D. H. Gustafson and W. F. Erman, Tetrahedron Lett. **5**, 401 (1967).
[7] Weissenborn, A., to VEB Wolfen, DDR 5051 (1952), DDR 8201 (1952).
[8] Kheifits, L. A., E. A. Simanovskaya, V. N. Belov, P. V. Ivanov, E. S. Shapiro and M. Y. Braines, Masloboino-Zhirovaya Prom. **23**, 35 (1957).
[9] Kheifits, L. A., G. I. Moldovanskaya, E. V. Broun and V. N. Belov, Zhur. Obshchei Khim. **30**, 1716 (1960).
[10] Kheifits, L. A., G. I. Moldovanskaya and L. M. Shulov, Zh. Org. Khim. **1**, 1057 (1965).
[11] Aul'chenko, I. S., and L. A. Kheifits, American Perfumer and Cosmetics **85**, 37 (1970).
[12] Erman, W. F., and T. J. Flautt, J. Org. Chem. **27**, 1526 (1962).
[13] Flautt, T. J., and W. F. Erman, J. Am. Chem. Soc. **85**, 3212 (1963).
[14] Erman, W. F., J. Am. Chem. Soc. **86**, 2887 (1964).
[15] Dahill, T. R. Jr., J. Dorsky and W. Easter, J. Org. Chem. **35**, 251 (1970).
[16] Dorsky, J., and W. M. Easter, Jr. (Givaudan Corp.), U.S. 3.499.937, patented Mar. 10, 1970.
[17] Kheifits, L. A., I. S. Aul'chenko and G. M. Shchegoleva, Zh. Org. Khim. **2**, 2059 (1966).
[18] Moldovanskaya, G. I., and L. M. Shulov, Zh. Org. Khim. **1**, 1057 (1965).
[19] Demole, E., Helv. Chim. Acta **52**, 2065 (1969).
[20] Stoll, M., France Parfums **8**, 227 (1965).
[21] Aul'chenko, I. S., and L. A. Kheifits, Probl. Poluch. Poluprod. Prom. Org. Sin., Akad. Nauk S.S.S.R., Otd. Obshch. Tekh. Khim. 122 (1967).
[22] Lange, G.-K., and K. Bauer, Patent pending.
[23] Clauß, K., and H. Jensen, Angew. Chem. **85**, 981 (1973).
[24] Muijlwijk van, A. W., A. P. G. Kieboom and H. van Bekkum, J. Roy. Neth. Chem. Soc. **93**, 204 (1974)..
[25] Vowinkel, E., and H.-J. Baese, Ber. **107**, 1213 (1974).
[26] Kenner, G. W., and N. R. Williams, J. Chem. Soc. 522 (1955).
[27] Haddad, Y. M. Y., H. B. Henbest, J. Husbands and T. R. B. Mitchell, Proc. Chem. Soc. 361 (1964).
[28] Krishnamurthy, S., and H. C. Brown, J. Am. Chem. Soc. **98**, 3383 (1976).

10. INFLUENCE OF THE POSITION OF THE METHYL GROUP UPON THE ODOR OF FATTY ALDEHYDES HAVING AN α- OR β-METHYL SUBSTITUENT

D. Hagena* and K. Bauer*

Summary α-methyl substituted fatty aldehydes with ten to thirteen carbon atoms are frequently used in perfumery because of their stability and their intense characteristic fragrances. The odors and application areas of these α-methyl aldehydes were compared to those of the corresponding β-methyl isomers, some of which were unknown in the literature. The lower β-methyl aldehydes, which are structurally somewhat similar to terpenes, display a distinct citrusy note which together with their softer and fuller fragrance give rise to completely new odor properties hitherto unknown for aliphatic aldehydes. The superior perfumistic properties of β-methyl aldehydes relative to their α-methyl isomers are demonstrated with examples, and methods for the synthesis of β-methyl aldehydes are described.

Introduction

Aliphatic aldehydes, particularly those with ten to thirteen carbon atoms (Fig. 1), are used extensively in perfumery, as they often give rise to piquant qualities when used in combination with other fragrances [1]. Unlike straight-chain aldehydes, α-methyl substituted aldehydes do not occur in nature [2]; yet, their complex fragrances, combining a characteristic fresh and slight fatty aldehyde note, result in their widespread use for producing certain top notes in perfumes [3]. The 2-methyl alkanals exhibit greater stability under slightly acidic or basic conditions than

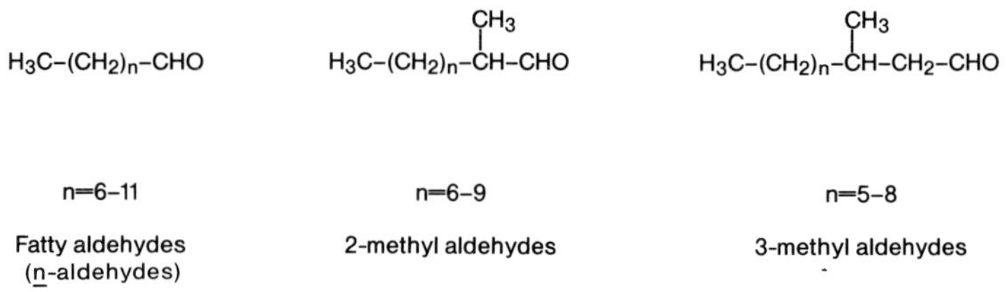

Fig. 1 Structures of *n*-aldehydes, 2-methyl aldehydes and 3-methyl aldehydes.

* D. Hagena and K. Bauer, Haarmann & Reimer GmbH, 3450 Holzminden, West Germany.

do the corresponding straight-chain aldehydes. This is because a labile α-hydrogen is replaced by a methyl group, thereby reducing the formation of condensation products and/or polymers such as trioxanes. Until recently, the most important 2-methyl aldehydes were those with eleven or twelve carbon atoms, namely methyl octyl acetaldehyde (2-methyl decanal) [4] and methyl nonyl acetaldehyde (2-methyl undecanal) [5–11]. Other α-substituted aldehydes appear only sporadically on the market and are of little importance.

Surprisingly, of the β-methyl substituted terpene-like aldehydes, only 3-methyl nonanal [12, 13, 14] and 3-methyl dodecanal [13, 15] were known. The corresponding 3-methyl aldehydes with eleven and twelve carbon atoms should also be of interest, but they apparently were not used in perfumery and had not been reported in the literature. These compounds have now been synthesized, and their fragrance properties investigated and compared with those of the 2-methyl aldehydes with ten to thirteen carbon atoms.

Fragrance Properties

The value to the perfumer of 2-methyl nonanal [16] (Fig. 2), the lowest member of the series, is very low. The odor impression is very simple, being somewhat metallic, chemical-like and exhibiting a fatty, green note. In contrast, 3-methyl nonanal [12, 13, 14] (Fig. 2) is a very valuable perfume chemical. The full, well-balanced fragrance character displays a certain citrus nuance and a dewy, calyx note. A lively, fresh and youthful top note is obtained through the use of this aldehyde.

Fig. 2 2-methyl nonanal | 3-methyl nonanal

2-methyl decanal [4] (Fig. 3) is sold by Haarmann & Reimer under the name "Oranginal", and it has a very intense fragrance somewhat reminiscent of bitter oranges, rue oil or savin oil. This aldehyde finds preferential application in citrus compositions, especially in the soap and washing powder fields where impact is very important. 3-Methyl decanal (Fig. 3), however, has a much more elegant fragrance which can be described as being a combination of a waxy, rosy citrusy note with a cyclamen-like, ozone, green note. This green note is very similar to that

Fig. 3 2-methyl decanal | 3-methyl decanal

typical for the peel of the South American lime. This aldehyde confers a well-balanced citrus-like freshness upon perfume compositions, and is thus ideally suited for Eaux de Cologne, citrus complexes, male lotions, fantasy and floral bases, cream perfumes and perfume oils for household products.

2-methyl undecanal [5–11] (Fig. 4), otherwise known as methyl nonyl acetaldehyde, is by far the most well-known and widely used of those branched chain aldehydes which do not occur in nature. It has a very popular, strong fragrance and is mostly used to reproduce conifer or, more specifically, silver fir notes. The very intense top note is fresh, woody and dry, somewhat reminiscent of incense and oranges. 3-methyl undecanal (Fig. 4), on the other hand, has a much more elegant fragrance and is more harmonious than the α-substituted isomer. The citrus-waxy note of this β-methyl aldehyde embodies very nicely the typical top note of mandarin orange peel, and this encourages its use in bases, perfumes, citrus complexes, cream perfumes and perfume oils for household products.

Fig. 4 2-methyl undecanal | 3-methyl undecanal

2-methyl dodecanal [17, 18] (Fig. 5) exhibits odor properties very similar to 2-methyl undecanal. However, because of its one-sided fatty, conifer-like nature, it finds very limited application in perfumes. By contrast, the fragrance impression imparted by 3-methyl dodecanal [13, 15] (Fig. 5), is much less boring due to its greater complexity. Many different aldehyde notes can be detected in this complex, but the dominating impression is that of the fresh, waxy fragrance of orange peel which now plays such an important role in modern perfumes with oriental notes.

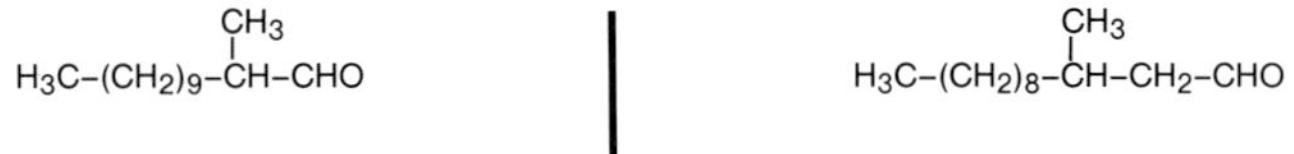

Fig. 5 2-methyl dodecanal | 3-methyl dodecanal

A comparison of the structures of 3,7-dimethyl octanal (tetrahydrocitral) and the 3-methyl alkanals reveals a distinct similarity between the latter and the terpenoid skeleton (Fig. 6). Thus, it is not too surprising that the 3-methyl substituted fatty aldehydes exhibit a terpenoid, citrus-like character. This citrus note is most intense in the case of 3-methyl nonanal, which shows the greatest degree of structural similarity to the terpenoid, and is, in fact, isomeric with 3,7-dimethyl octanal. The odor similarity between tetrahydrocitral and the 3-methyl aldehydes diminishes with increasing carbon chain length of the methyl substituted alkanal; however, the citrus note does remain as a characteristic feature. One of the most conspicuous aspects of the

3-methyl aldehydes is their softer and fuller fragrance relative to their 2-methyl analogues. They elicit an extremely harmonious impression, displaying, in addition to their terpenoid character, a note quite characteristic of fatty aldehydes. Thus, the 3-methyl aldehydes give rise to completely new odor properties hitherto unknown for fatty aldehydes.

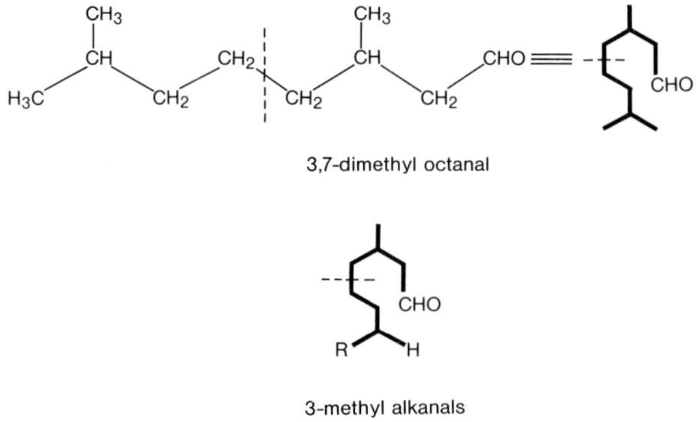

Fig. 6 Comparison of the structure of the terpenoid 3,7-dimethyl octanal with those of 3-methyl alkanals.

Syntheses of 3-Methyl Aldehydes

In contrast to the 2-methyl aldehydes, the 3-methyl isomers which were previously known were hardly ever used in perfumery. This is probably due to the fact that, unlike the synthesis of 2-methyl aldehydes, very few suitable reactions are available for the synthesis of the terpene-like 3-methyl branched aldehydes. Such procedures, as are available, are more complex and consequently more expensive than those for the 2-methyl isomers.

A synthesis of 3-methyl substituted aldehydes which might be economically viable consists of reacting esters of crotonic acid with appropriate alkyl halides under Grignard reaction conditions [19] (Fig. 7). However, good yields can only be obtained when esters of secondary

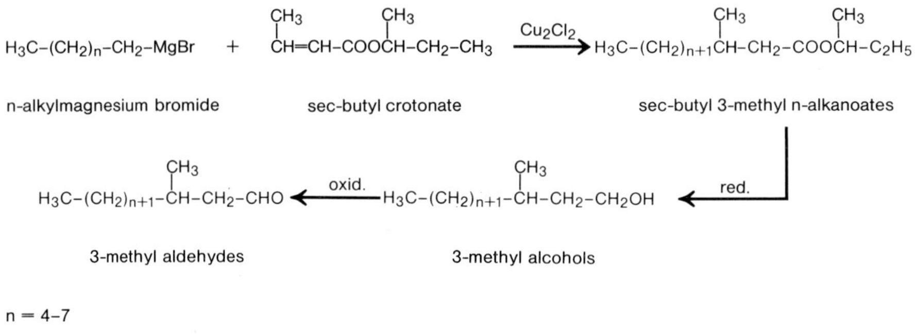

Fig. 7 Synthesis of 3-methyl aldehydes via the reaction of crotonate esters with *n*-alkyl magnesium bromides.

alcohols are used, since otherwise 1,2-addition occurs. Another synthesis described in the literature [12] involves the ozonolysis of 4-methyl dec-1-ene (Fig. 8), but this synthesis has no practical importance because the starting material is very difficult to obtain.

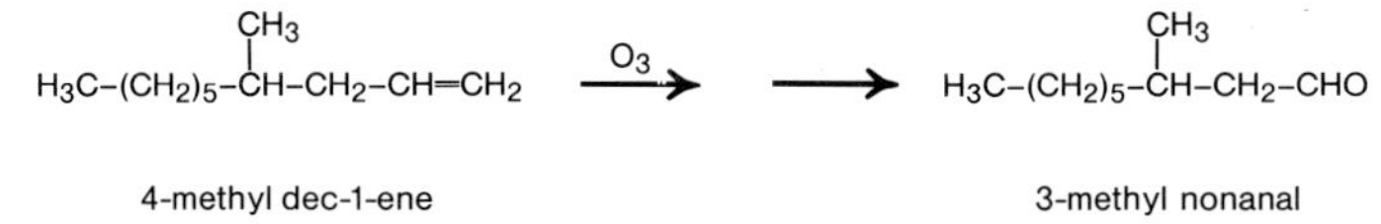

Fig. 8 Synthesis of 3-methyl nonanal via ozonolysis of 4-methyl dec-1-ene.

Since the methods available for the synthesis of 3-methyl aldehydes were unsatisfactory for several reasons, an economical procedure for the synthesis of 3-methyl alkanals was needed. A three step reaction which allows 3-methyl aldehydes to be obtained in high yields was subsequently developed [20, 21]. The first step of the procedure is illustrated in Figure 9. When

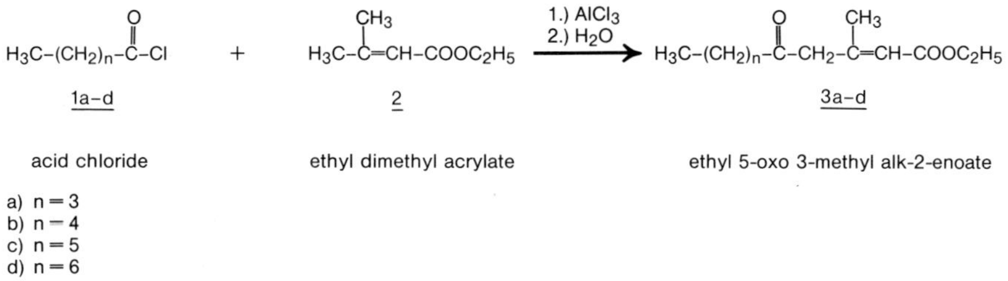

Fig. 9 First step in a new economical synthesis of 3-methyl aldehydes involving reaction of an acid chloride (*1*) with ethyl dimethyl acrylate (*2*) under Friedel-Crafts conditions to afford ethyl 5-oxo 3-methyl alk-2-enoate (*3*).

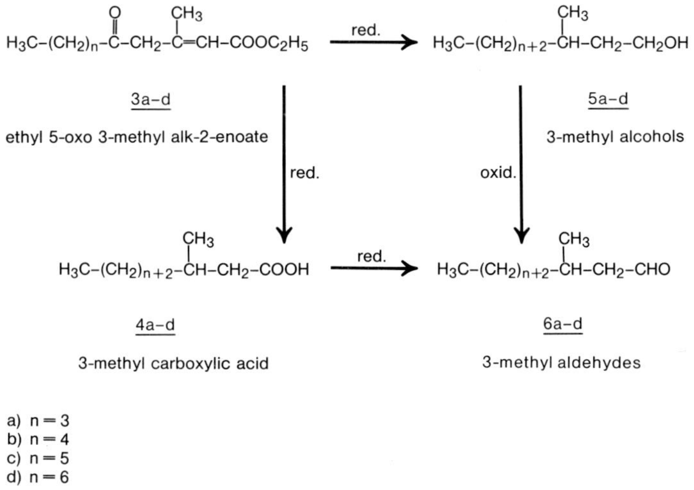

Fig. 10 Catalytic reduction of the unsaturates keto ester (*3*) to the corresponding acid (*4*) or alcohol (*5*), both of which may be converted to the corresponding 3-methyl aldehyde (*6*).

the ethyl ester of dimethyl acrylic acid (*2*) is reacted with an acid chloride (*1*) under Friedel-Crafts conditions, acylation of one of the methyl groups takes place, and after hydrolytic work-up the unsaturated keto ester (*3*) is obtained. The unsaturated keto esters (*3 a–d*) can then be converted into the corresponding acids (*4 a–d*) or alcohols (*5 a–d*) in a single catalytic reduction step (Fig. 10). These acids or alcohols are subsequently transformed into the corresponding 3-methyl aldehydes via either an additional reduction or oxidation step, respectively. Thus, these new and extremely interesting products, which exhibit a high degree of fragrance acceptability, are now available to perfumers.

Acknowledgement

We would like to express our thanks to H. O. Müller for carrying out all the perfumistic evaluations, and to Drs. Mack and Köpsel for the analytical aspects of the research.

References

[1] Dorland, W. E., J. A. Rogers Jr., The Fragrance and Flavor Industry S. 144, Dorland, W. E., Company Box 264, Mendham, New Jersey 07945 (1977).
[2] Bedoukian, P. Z., Perfumery Synthetics and Isolates S. 33, D. Van Nostrand Company, Inc. Toronto, New York, London (1951).
[3] West, T. F., H. J. West, H. J. Strausz, D. H. R. Barton, Synthetic Perfumes p. 90, Edward Arnold + Co., London (1949).
[4] Arctander, St., Perfume and Flavor Chemicals, Vol. I/II No. 2145; St. Arctander, Montclair, New York 1969.
[5] Clemente, I., Riechstoffind. **2**, 77 (1927).
[6] Morel, C., Soap Perfumery and Cosmetics **25**, 512 (1952).
[7] Unilever N. V., NL 6.503.576 (1964), through Chem. Abstr. **64**, 4860 d (1966).
[8] Land, D. G., R. Harper, M. Griffiths, Flavour Ind. **1** (12), 842 (1970).
[9] Curtis, A. J., B. G. Jaggers, F. Janes, to Bush Boake Allen Ltd., Ger. Offen. 1.948.536 (1968), through Chem. Abstr. **74**, 91095 b (1971).
[10] Voitkevich, S. A., E. I. Gushchina, T. A. Rudolfi, M. A. Nikitina, O. G. Dokunina, A. G. Belfer, U.S.S.R. 465941 (1976), through Chem. Abstr. **85**, 166491 r (1976).
[11] Lit. 4, No. 2132.
[12] Letsinger, R. L., J. G. Traynham, J. Am. Chem. Soc. **72**, 849 (1950).
[13] Karrer, P., Lehrbuch der organischen Chemie p. 190, Georg Thieme Verlag Stuttgart (1963).
[14] Lit. 4, No. 2136.
[15] Lit. 4, No. 2141.
[16] Lit. 4, No. 2047.
[17] Cason, J., K. L. Rinehart, Jr., J. org. Chem. **20**, 1591 (1955).
[18] Lit. 4, No. 1997.
[19] Munch-Petersen, T., J. Org. Chem. **22**, 170 (1957).
[20] Alkonyi, J., Chem. Ber. **98**, 3099 (1965).
[21] Lohaus, G., G. W. Friedrich, J. P. Jeschke, Chem. Ber. **100**, 658 (1967).

11. OLD AND NEW MACROCYCLIC MUSKS: A GENERAL REVIEW FROM THE INDUSTRIAL STANDPOINT

K. A. Bauer* and A. K. Körber*

Summary The 50 year old history of the industrial manufacture of macrocyclic musks provides a good example of the rapid utilization of recent scientific findings by industrial applied research departments. Yet, as demonstrated by comparison of the well known literature compounds with newly discovered musks, there are present limitations in certain areas of the industrial preparative chemistry of these macrocyclic compounds. Nevertheless, the development of more effective procedures for the manufacture of these extremely valuable perfume compounds will remain one of the main areas of future research. This is especially true for the preparation of macrocyclic esters and lactones because, of the many different types of musks, these types largely satisfy product safety requirements.

Introduction

The discovery and structural elucidation of the natural macrocyclic musks, culminating in their synthesis on an industrial scale, is a long and fascinating history that has been covered in much detail elsewhere [1]. It suffices to point out here that, judging by its prominent mention in the Koran, musk was, historically, one of the most valuable of all aromas. The macrocyclic musks are still among the most desirable of fragrances, and clearly among the most costly.

The traditional musk is, of course, an animal secretion, yet a few C_{15}-C_{17} macrocyclic musks with a lactone structure have been isolated from plants. The classical example is pentadecanolide, first isolated from angelica root oil [2]. On the other hand, the number of C_{15}-C_{17} macrocyclic ketones which have been detected in the secretions of the musk deer, civet cat, musk rat and other animal species is steadily increasing [3]. A number of naturally occurring macrocyclic musks with a lactone or ketone structure are listed in Table 1.

Industrial Synthesis of Macrocyclic Musks

During the past few years, numerous syntheses have been developed for both macrocyclic lactones and macrocyclic ketones, but few have actually found industrial application. The main

* K. A. Bauer and A. K. Körber, Haarmann & Reimer GmbH, 3450 Holzminden, West Germany.

Table 1

Naturally Occuring Macrocyclic Musks

Ketones		Lactones	
a) straight chain		a) straight chain	
$C_{15}H_{28}O$	cyclopentadecanone	$C_{15}H_{28}O_2$	15-pentadecanolide
$C_{15}H_{26}O$	cis-cyclopentadec-5-en-1-one	$C_{16}H_{30}O_2$	16-hexadecanolide
$C_{15}H_{24}O$	cyclopentadec-5-yn-1-one	$C_{16}H_{28}O_2$	16-hexadec-7-enolide = ambrettolide
$C_{16}H_{30}O$	cyclohexadecanone		
$C_{17}H_{32}O$	cycloheptadecanone		
$C_{17}H_{30}O$	cis-cycloheptadec-5-en-1-one	b) branched chain	
$C_{17}H_{30}O$	cis-cycloheptadec-6-en-1-one	$C_{15}H_{28}O_2$	14-pentadecanolide
$C_{17}H_{30}O$	cis-cycloheptadec-7-en-1-one	$C_{16}H_{30}O_2$	15-hexadecanolide
$C_{17}H_{30}O$	cis-cycloheptadec-9-en-1-one		
$C_{17}H_{28}O$	cis,cis-cycloheptadeca-5,11-dien-1-one		
$C_{17}H_{28}O$	cycloheptadec-5-yn-1-one		
$C_{17}H_{26}O$	cis-cycloheptadec-7-en-5-yn-1-one		
b) branched chain			
$C_{16}H_{30}O$	3-methylcyclopentadecanone = muscone		

reasons for this are the large number of reaction steps necessary for the synthesis of the macrocycle and the poor yields obtained with some of the older ring closure reactions. It should be mentioned in this context that, during the past seventy to eighty years, numerous musks have been developed which are structurally unrelated to the natural musks [4]. These compounds were considerably easier to synthesize and were much cheaper than the natural macrocyclic product. Furthermore, they have proved themselves to be satisfactory replacements for the natural product. Thus, it was only after the discovery of the relationship between the macrocyclic structure and musk odor [5], that macrocyclic lactones and ketones which did *not* occur in nature were synthesized. These compounds display an intense musk odor when certain conditions are fulfilled (e.g., the number of atoms in the ring should be between 15 and 17), and possess superior odor properties compared to the previously mentioned synthetic substitutes. Yet, only a few macrocyclic musks, namely those which could be produced simply and cheaply, actually appeared in large quantities on the market.

Ring closure is an important step in the synthesis of many macrocyclic musks, and is best achieved by applying the Carothers depolymerization technique [6] to suitable polyesters, which are obtained relatively easily from ω-hydroxy acids, or dicarboxylic acids and glycols. Thus, it is hardly surprising that all the newly introduced synthetic macrocyclic musks were lactones. The main synthetic problem with these lactones was not whether ring closure could be effected, but rather whether a viable synthesis of the ω-hydroxy acid or the dicarboxylic acid could be achieved. The preparation of the starting materials for the synthesis of macrocyclic musks therefore received considerable attention. Thus, brassylic acid (1,13-tridecandioic acid), which is easily obtained via the ozonolysis of erucic acid, was utilized [7] for the preparation of ethylene brassylate (ethylene 1,13-tridecandioate) (Fig. 1). Similarly, ω-hydroxy-12-oxa-hexadecanoic acid, needed for the synthesis of 12-oxa-hexadecanolide (Fig. 1), is the product of the condensation reaction between ω-bromoundecanoic acid and 1,4-butylene glycol [8]. Likewise, once 1,12-dodecanedioic acid was established as a technical product [9], there soon followed the ethylene ester (Fig. 1) which displays a musk odor and is easily synthesized [7].

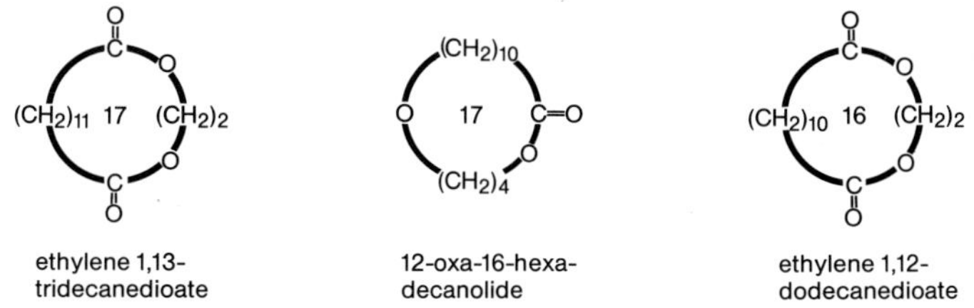

Fig. 1 Structures of ethylene brassylate (ethylene 1,13-tridecanedioate) and related macrocyclic lactones.

The industrial manufacture of macrocyclic musks gathered new momentum as a result of the ready availability of cheap cyclododecane derivatives [10]. Thus, instead of ring closure reactions, ring expansion reactions achieved prominence and in this way macrocyclic ketones were easily obtained. As a result, cyclohexadec-5-en-1-one (Fig. 2), which can be synthesized in a few steps from cyclododecanone [11], appeared in large quantities on the market. Using the same cyclododecanone as starting material, an industrially viable procedure was also developed for the manufacture of the lactone, 15-pentadecanolide (Fig. 3, see also symposium paper No. 12). This lactone is the only naturally occurring musk which, due to its exceptional odor properties, had attained considerable industrial importance in earlier times.

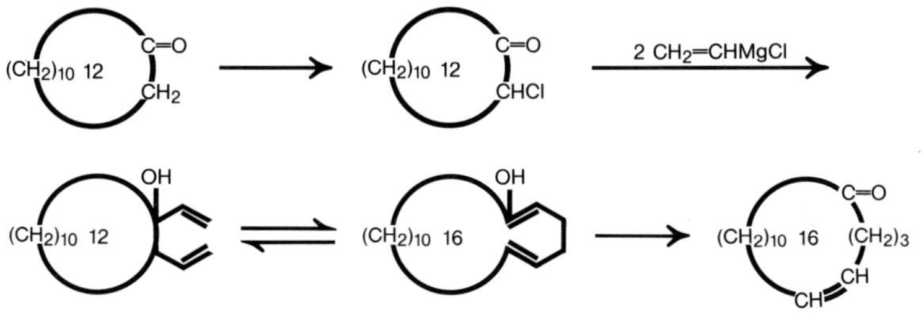

Fig. 2 Scheme for the preparation of cyclohexadec-5-en-1-one from cyclododecanone.

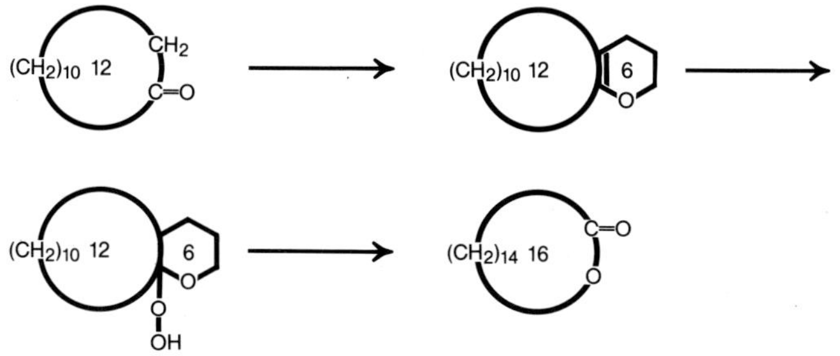

Fig. 3 Scheme for the preparation of 15-pentadecanolide from cyclododecanone. For details of the synthesis of 15-pentadecanolide see paper no. 12.

New Synthetic Approaches and New Musks

While a number of synthetic macrocyclic musks have found commercial application, the high production costs for some of the perfumers more desirable musks have prevented their firm establishment in the marketplace. Thus, new and economical processes using different raw materials or methods were needed to manufacture these valuable compounds. In this context, we focused our attention on 15-pentadecanolide and muscone for a number of reasons, including their natural occurrence, their fragrance intensity, their high fragrance qualities and their characteristic notes. The results of this work are described in papers No. 12 and No. 13 of this symposium and will not be reviewed here. Rather, I wish to expand on our second objective, which was to synthesize entirely new musks by exploiting the known relationship between fragrance and chemical structure, and to thereby produce musks superior to those now available.

At the beginning of our systematic investigations, we focused specifically on dilactone musks since these compounds can be obtained very easily from diols and dicarboxylic acids. While ethylene brassylate (ethylene 1,13-tridecanedioate) possesses a relatively intense odor, it was surprising that there were no data in the literature on the seventeen-membered ring isomeric dilactone obtained from oxalic acid and tridecane-1,13-diol (i.e., tridecamethylene oxalate) (Fig. 4). As it transpired, the odor of tridecamethylene oxalate was even more intense than that of ethylene brassylate, even though the oxalate had a lower volatility [12]. In fact, even today a strong musk odor can be detected on a smelling blotter which was moistened with this dilactone over three years ago. Thus, relatively simple structural modification of a known musk resulted in an even more intense musk. The fact that tridecamethylene oxalate had not been reported previously could be due to the fact that it is considerably less stable than ethylene brassylate under both acidic and alkaline conditions, and this automatically limits its use as a musk odorant.

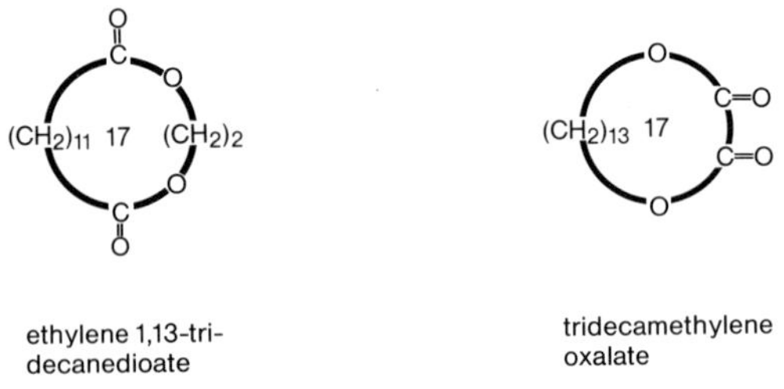

Fig. 4 Comparison of the structures of ethylene brassylate (ethylene 1,13-tridecanedioate) and the isomeric dilactone tridecamethylene oxalate.

In the case of the macrocyclic ketones and monolactones, it is possible to replace a methylene group by an oxygen atom without losing much of the musk odor [13]. This is also the case with the dilactones. Thus, 7-oxabrassylic acid, which can be obtained from caprolactone, can be transformed into its ethylene ester in a manner analogous to that of brassylic acid (Fig. 5). This new ethylene ester, along with the isomeric oxalate of dihexamethylene glycol (Fig. 5), constitute the first representatives of a hitherto unknown series of macrocyclic musks with an oxa-dilactone structure [14]. Of all the seventeen-membered ring musks, ethylene 7-oxa-1,13-tridecanedioate and 7-oxatridecamethylene oxalate display the highest water solubilities. However, the third ring oxygen atom has the expected negative effect upon the odor of these compounds. As before, the oxalate has the more intense odor of the two isomers (Fig. 5). The cyclic esters which could be synthesized from 5- or 6-oxabrassylic acid and glycol, or from the corresponding diols and oxalic acid, might possibly have an even more intense odor. The preparation and testing of these compounds could make an important contribution towards understanding the very complex structure-odor relationships of macrocyclic musks, a topic recently reviewed by Beets [15] who also discusses the difficulties encountered in evaluating musk odors.

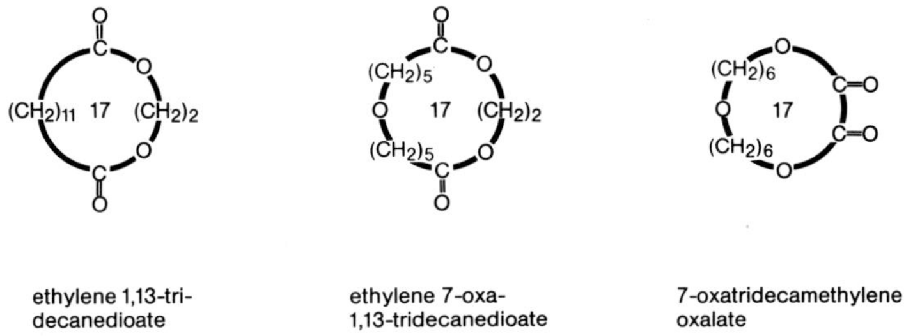

Fig. 5 Ethylene brassylate and some related 7-oxa-dilactones.

In the course of our research, a procedure was developed for the synthesis of β-methyl substituted dicarboxylic acids (see symposium paper No. 13). It was therefore a natural extension of this work to prepare the corresponding macrocyclic esters, which appeared to be completely unknown, and to compare the properties of these compounds with those of the isomeric ethylene brassylate and its homologues (Fig. 6). Surprisingly, the presence of the methyl group causes the ethylene esters of β-methyl dodecanedioic acid and β-methyl brassylic acid to display a secondary odor note in addition to the expected musk note. The odors of these compounds show a remarkable resemblance to costus root oil, thus providing entirely new and exciting possibilities for the perfumer.

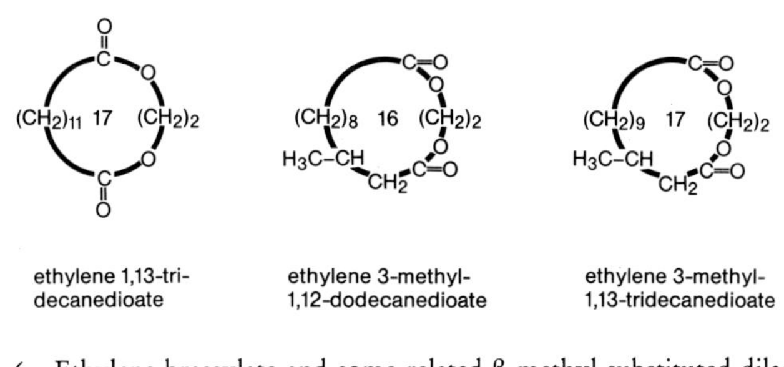

Fig. 6 Ethylene brassylate and some related β-methyl substituted dilactones.

As a final note, the macrocyclic ketone 3-methylcyclohexadec-5-en-1-one (Fig. 7) deserves mention. This ketone is an isomer of civetone and it could also be considered an unsaturated homomuscone; yet, it might best be described as a hybrid of muscone and civetone which has inherited only the positive attributes of both parents. The fragrance properties of this compound are among the very best of the macrocyclic musks, being very similar to those of muscone. On the other hand, its fixative properties resemble those of civetone. The synthesis and properties of 3-methylcyclohexadec-5-en-1-one and other interesting macrocyclic ketones are described in symposium papers No. 13 and No. 14.

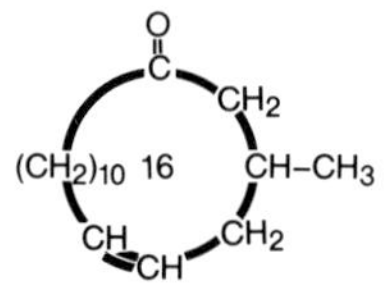

3-methylcyclohexadec-5-en-1-one

Fig. 7 3-Methylcyclohexadec-5-en-1-one.

Concluding Remarks

While confirming the well-known effect that certain parameters modify the properties of macrocyclic musks, this short review of one of the most fascinating areas of industrial chemistry has probably raised more questions than it has answered. Surely one of the most interesting questions is whether the macrocyclic musks would have attained their present status if the civet cat and musk deer had *not* existed a few thousand years ago. This question is clearly unanswerable. On the other hand, the question of whether synthetic materials will ever surpass the natural animal pheromones, as regards their perfumistic and technological qualities, can probably be answered in the affirmative, provided that synthetic compounds structurally related to the natural musks are considered.

Acknowledgement

Industrial research is predominantly applied research, the magnitude and scope of which are largely determined by the likelihood of obtaining results capable of commercial exploitation. Considering the relatively wide range of acceptable macrocyclic musks, and the even larger number of different methods for their synthesis, our attempt to enter this field was fraught with risks. The fact that we were successful is entirely due to the perseverance and high degree of commitment shown by those colleagues who were involved in this research. We would especially like to express our thanks and gratitude to Drs. Hopp and Hagena.

References

[1] Wood, T. F., Givaudanian **6,** 6 (1975).
[2] Kerschbaum, M., Ber. **60B,** 902 (1927).
[3] Van Dorp, D. A., R. Klok and D. H. Nugteren, Recueil **92,** 915 (1972).
[4] Wood, T. F., Chemistry of the Aromatic Musks, Givaudanian, separate print.
[5] Ruzicka, L., Helv. **9,** 230, 715, 1008 (1926).
[6] Hill, J. W., and W. H. Carothers, J. Am. Chem. Soc. **55,** 5031–5039 (1933).
[7] Spanagel, E. W., and W. H. Carothers, J. Am. Chem. Soc. **57,** 929–934 (1935).
[8] Spanagel, E. W., and W. H. Carothers, J. Am. Chem. Soc. **58,** 654 (1936).
[9] Roehl, H. W. Eversmann, P. Hegenberg and G. Hellemanns (to Chemische Werke Huels AG), DOS 2 001 182 (1969).
[10] Wilke, G., Angew. Chem. **69,** 397 (1957).
[11] Nishino, M., H. Kondo and A. Miyake, Chem. Letters **1973,** 667.
[12] Unpublished.
[13] Berends, W., Parf. Cosm. Sav. **8** (11), 475 (1965).
[14] Unpublished.
[15] Beets, M. G. J., Structure-Activity Relationship in Human Chemoreception, Applied Science Publishers Ltd., London (1970).

12. NEW WAYS FOR THE SYNTHESIS OF 15-PENTADECANOLIDE

A. K. Körber* and K. A. Bauer*

Summary Pentadecanolide, which was first discovered by M. Kerschbaum in angelica root oil, plays a dominant role among the naturally occurring macrocyclic musk odorants. An overview of the known synthetic methods is followed by a description of the new Haarmann & Reimer procedure used for the manufacture of this lactone. Thus, cyclododecanone is reacted with allyl alcohol to yield 13-oxabicyclo[10.4.0]-hexadeca-1(12)-ene (*1*) which, in turn, is efficiently converted to the macrocyclic lactone in three further steps.

Scheme 1

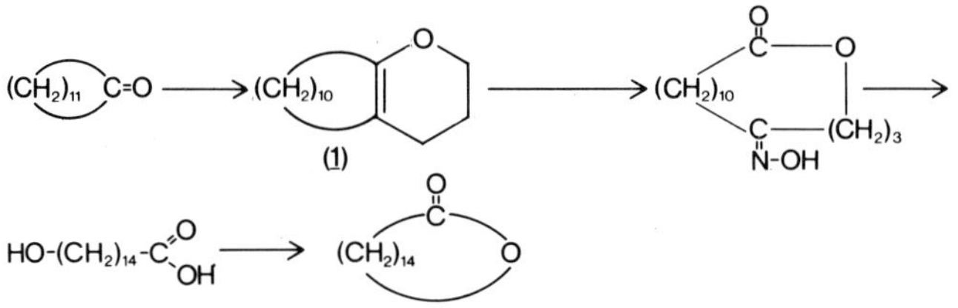

Introduction

Pentadecanolide is a naturally occurring musk odorant which was first discovered in angelica root oil [1]. The chemical structure of 15-pentadecanolide was elucidated in our research laboratories by Max Kerschbaum, who also carried out the first synthesis of this substance. The method was published in March, 1926 in the form of a patent entitled "A Procedure for the Synthesis of Lactones with a Musk Odor" [2]. Although many musk odorants have been synthesized and have become commercially available, pentadecanolide is of special interest for the perfumer because of its intense but fine musk odor and its excellent fixative properties. From the moment it was discovered, there has been continual interest in developing technically feasible syntheses which would make this macrocyclic lactone generally available to the perfume industry.

---

* A. K. Körber and K. A. Bauer, Haarmann & Reimer GmbH, 3450 Holzminden, West Germany.

Early Synthetic Approaches to 15-Pentadecanolide

The very first synthesis of 15-pentadecanolide was that of M. Kerschbaum, who caused the silver salt of 15-bromopentadecanoic acid to cyclize thereby yielding the lactone [2] (Fig. 1).

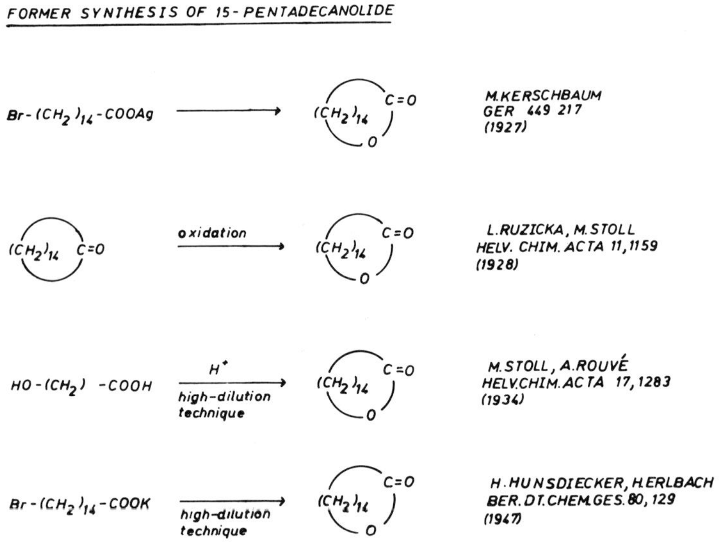

Fig. 1 Some earlier syntheses of 15-pentadecanolide.

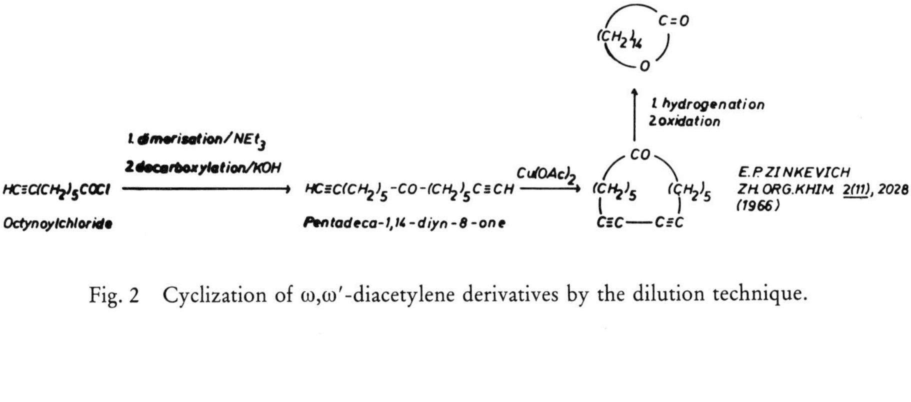

Fig. 2 Cyclization of ω,ω′-diacetylene derivatives by the dilution technique.

This was followed by the procedure of Ruzicka and Stoll [3], who made use of the Bayer-Villiger oxidation of ketones using Caro's acid. Thus, pentadecanolide was synthesized from the corresponding cyclopentadecanone. Further progress was made by Stoll and Rouvé [4] who successfully applied the Ruggli-Ziegler dilution principle to the cyclization of 15-hydroxypentadecanoic acid [5] (Fig. 1). Although the dilution technique showed distinct advantages over the methods described earlier, it remained of academic interest only since very poor space/time yields were obtained. Figure 2 illustrates two syntheses based on the copper acetate-mediated cyclization of ω,ω'-diacetylene derivatives [6, 7]. These procedures have been carried out on a small scale only, due to the high degree of dilution that is necessary for the cyclization step.

The first genuine breakthrough in this field came as a result of the brilliant work of Carothers and co-workers who investigated the synthesis of macrocyclic lactones via depolymerization [8]. In this process (Fig. 3), the polyesters of ω-hydroxycarboxylic acids are converted into the monomeric macrocyclic lactones in the presence of suitable transesterification catalysts. High temperatures and a good vacuum are necessary in order to obtain good yields in the range of 80–90%. Thus, a cyclization procedure was now available which was suitable for the production of pentadecanolide on an industrial scale. Gradually other patents appeared which were based on Carothers' work and which claimed that esters, other than polyesters, of ω-hydroxy-carboxylic acid could be utilized for lactone formation. For example Givaudan [9, 10] used the glycerol ester, while Soda Koryo [11] utilized esters of low boiling alcohols (Fig. 3). Additionally, the Hasegawa firm [12] developed other catalysts, such as aluminium alcoholates, for the depolymerization step.

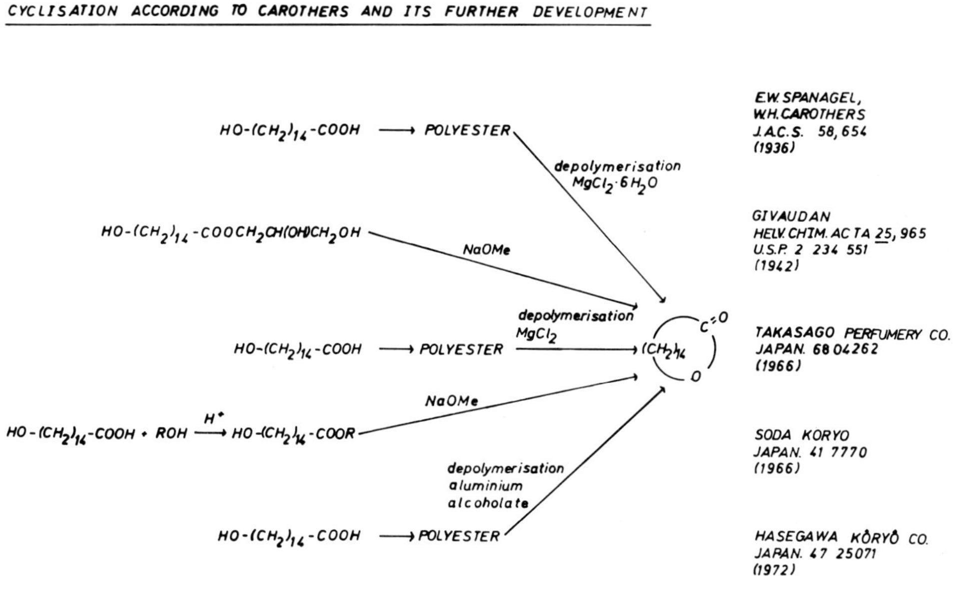

Fig. 3 Cyclization according to Carothers [8] and further development of the technique.

The relatively recent advent of cyclododecanone as an industrial product [13] prompted the development of entirely new strategies for the large scale synthesis of pentadecanolide, and a synthetic route based on this intermediate was discovered independently by both Haarmann & Reimer [14] and Firmenich [15] (Fig. 4). The key compound in this synthesis is the bicyclic ether 13-oxabicyclo-[10.4.0]hexadec-1(12)-ene (commonly known as decamethylenedihydropyran) which can easily be obtained from cyclododecanone as described later. Decamethylenedihydropyran is treated with hydrogen peroxide to form the saturated 12-hydroperoxy compound. When this bicyclic hydroperoxide undergoes thermal decomposition in the presence of a solvent which scavenges oxygen radicals, a mixture of macrocyclic lactones is obtained of which 15-pentadecanolide is the main product. However, this very interesting ring expansion reaction does display certain disadvantages such as unsatisfactory fragmentation yields and, as is often the case with radical reactions, a complex mixture of reaction products. In addition to the required 15-pentadecanolide, the product mixture contains 12-hydroxypentadecanolide, isomeric pentadecenolides, a dimeric peroxy compound, and a number of compounds originating from the solvent (Fig. 4). Only by submitting the reaction mixture to a complicated work up can pentadecanolide be obtained in a pure, perfumistically acceptable form. In summary, two practical methods have evolved as a result of the search for suitable procedures for the synthesis of pentadecanolide. These two methods are: [1], the synthesis of 15-hydroxypentadecanoic acid followed by condensation and depolymerization according to Carothers [8]; and [2], the conversion of cyclododecanone into a bicyclic derivative which undergoes subsequent expansion to form the lactone [16].

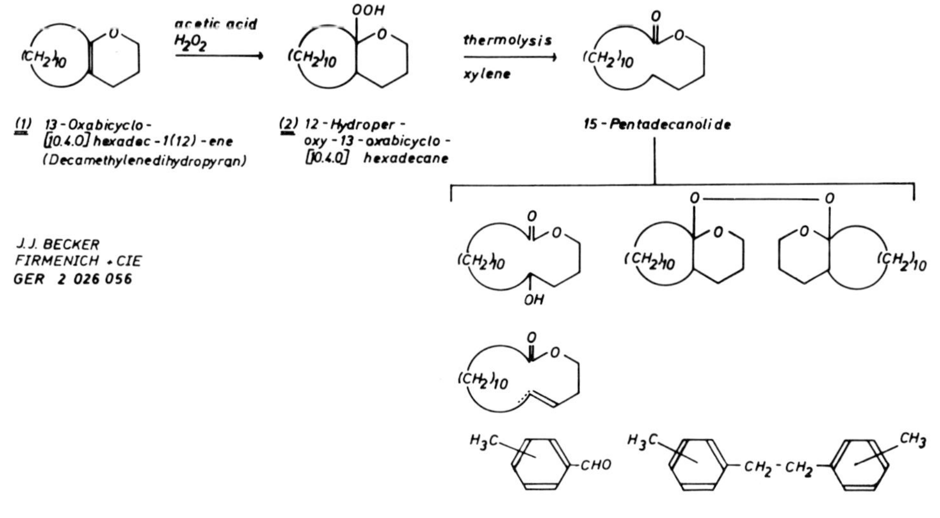

Fig. 4 A synthesis of 15-pentadecanolide from decamethylenedihydropyran (*1*).

The Haarmann & Reimer Synthesis of 15-Pentadecanolide

During the course of our research we were successful in combining the very good yields obtained by cyclization of 15-hydroxypentadecanoic acid with the economically favorable raw material, cyclododecanone and thereby developed an optimal, industrially feasible process for the synthesis of pentadecanolide [18]. In this procedure, cyclododecanone is converted in two steps into the previously described decamethylenedihydropyran (Fig. 5). Thus, cyclododecanone undergoes a radical reaction with allyl alcohol in the presence of di-*t*-butylperoxide as catalyst. The product of this reaction, 3-(2-oxocyclododecyl)-propan-1-ol (*3*), is subsequently cyclized and dehydrated to the bicyclic enol ether decamethylenedihydropyran [14]. A related procedure patented by Firmenich [15] requires five steps to synthesize this compound (Fig. 5, bottom).

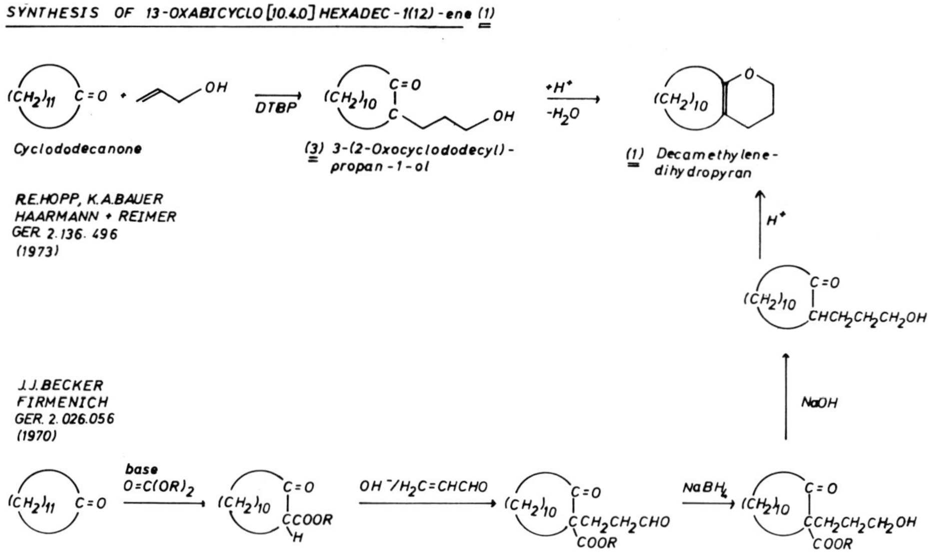

Fig. 5 Syntheses of decamethylenedihydropyran (*1*) from cyclododecanone.

In the second stage of the Haarmann & Reimer process, decamethylenedihydropyran undergoes ionic cleavage in the presence of nitric acid to form a sixteen-membered ring hydroxyimino-lactone (Fig. 6). This reaction, which was first reported by Mahajan and

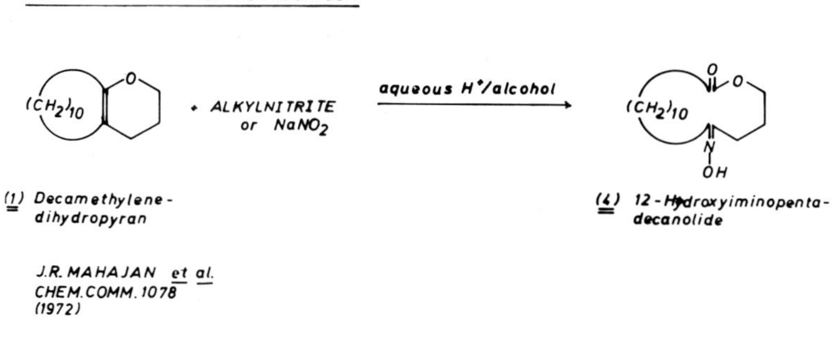

Fig. 6 Ring expansion of decamethylenedihydropyran (*1*) via nitrosation.

co-workers [17], can be carried out such that nearly quantitative yield of 12-hydroxyimino-15-pentadecanolide (*4*) is obtained. The 12-hydroxyimino-15-pentadecanolide (*4*), which has a weak musk odor, undergoes facile hydrolysis with sodium bisulfite to afford 12-oxo-15-pentadecanolide (*5*) in very high yield (Fig. 7). This latter compound (*5*), which also

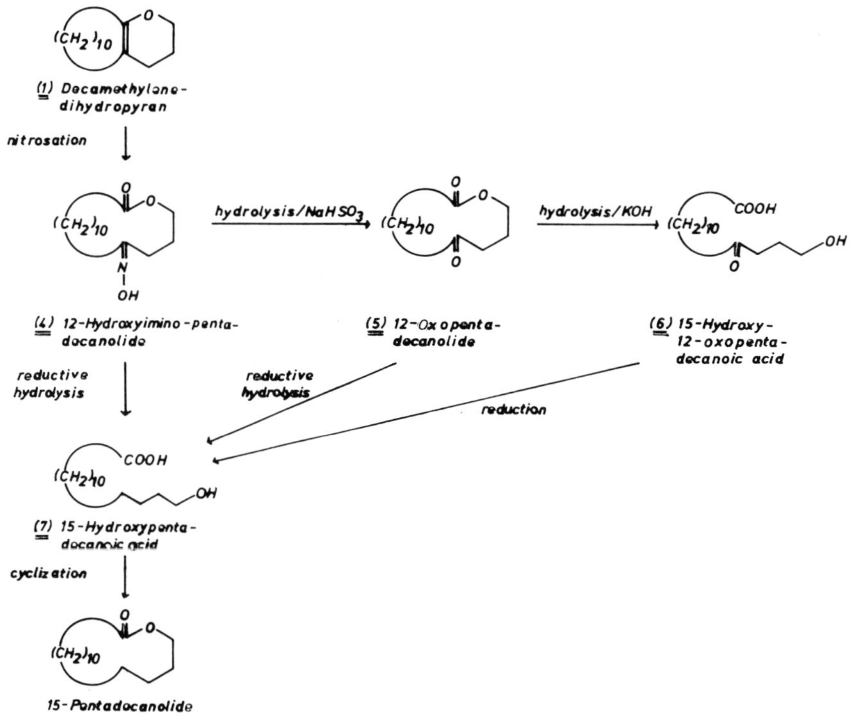

Fig. 7 Synthetic routes from decamethylenedihydropyran [*1*] to 15-pentadecanolide.

displays a weak musk odor, is then hydrolyzed under alkaline conditions to yield the 15-hydroxy-12-oxopentadecanoic acid (*6*), which is subsequently reduced under classical Wolff-Kishner reaction conditions to the corresponding 15-hydroxypentadecanoic acid (*7*). As indicated in Figure 7, the 12-hydroxyimino-15-pentadecanolide (*4*) can also be converted to the 15-hydroxypentadecanoic acid in one step. The reaction is carried out with hydrazine under alkaline conditions and yields the 15-hydroxypentadecanoic acid (*7*) in an excellent overall yield of 95%. The surprisingly good yields obtained for this one step reductive hydrolysis may be explained by the fact that hydroxylamine, which is a product of the reaction, decomposes with the formation of ammonia and nitrogen, and that the intermediate hydrazine, which is not isolated, decomposes with the liberation of nitrogen (Fig. 8). Thus, these two gas forming reactions may cause the equilibrium to shift, thereby greatly favoring the formation of the 15-hydroxypentadecanoic acid. It should be noted, at this point, that Zakharkin and co-workers [19] attempted to synthesize 15-pentadecanolide by the Clemmensen reduction of 12-oxo-15-pentadecanolide, but very poor yields were obtained.

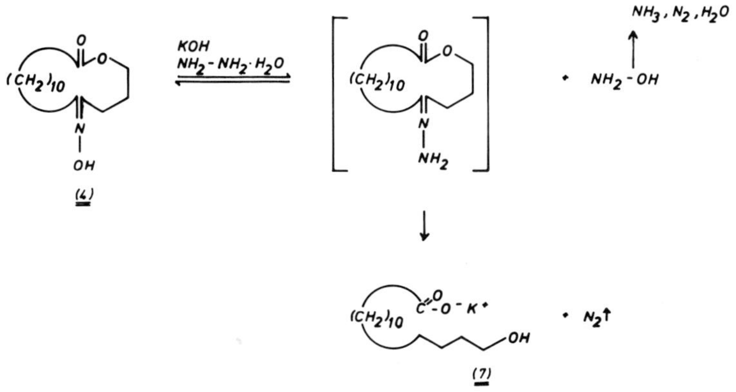

Fig. 8 Reductive hydrolysis of 12-hydroxyimino-15-pentadecanolide (*4*) to 15-hydroxypentadecanoic acid (*7*).

The process described above yields the 15-hydroxypentadecanoic acid in an extremely pure state, and we discovered later that this high degree of purity is of utmost importance for the subsequent polycondensation–depolymerization reaction. Thus, when using this acid, 15-pentadecanolide is obtained with yields in the region of 85–90% and with a purity of some 98% as determined gas chromatographically. A simple fractional distillation is all that is necessary to produce a perfume grade product. The new Haarmann & Reimer process for the synthesis of pentadecanolide is shown schematically in Figure 9.

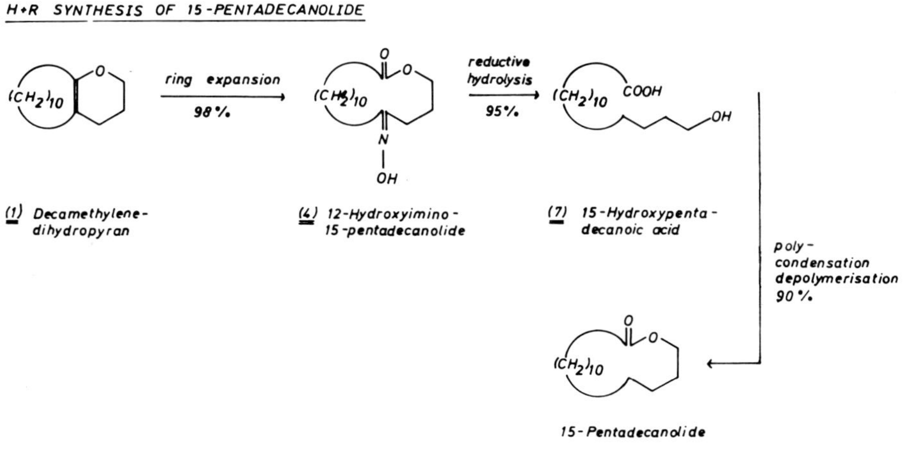

Fig. 9 The Haarmann & Reimer Synthesis of 15-pentadecanolide.

At first sight, it certainly seems odd that, in order to synthesize this macrocyclic product, the preformed ring has to be opened and then closed again. Needless to say, we have performed numerous experiments designed to avoid cleavage of the preformed lactone bond. Our results indicate, however, that the detour via the ring opening reaction is, in fact, advantageous because of the almost quantitative fragmentation of the enol ether and the excellent yield of the reductive cleavage step, and because of the fact that the very pure 15-hydroxypentadecanoic acid obtained can be converted in very high yield to pentadecanolide. As a result of this new process, Haarmann & Reimer is now in a position to offer pentadecanolide of the highest analytical purity and olfactory quality to meet the ever increasing needs of the perfume market.

References

[1] Kerschbaum, M., Ber. dtsch. chem. Ges. **60**, 902 (1927).
[2] Kerschbaum, M., Haarmann & Reimer, Holzminden, DRP 449217 (1927).
[3] Ruzicka, L., M. Stoll, Helv. Chim. Acta **11**, 1159 (1928).
[4] Stoll, M., A. Rouvé, Helv. Chim. Acta **17**, 1283 (1934).
[5] Hunsdiecker, H., H. Erlbach, Ber. dtsch. chem. Ges. **80**, 129 (1947).
[6] McCrae, W., Tetrahedron **20**, 1773 (1964).
[7] Zinkevich, E. P., Zh. Org. Khim. **2** (11), 2028 (1966).
[8] Spanagel, E. W., W. H. Carothers, J. Am. Chem. Soc. **58**, 654 (1936).
[9] Collaud, C., Helv. Chim. Acta **25**, 965 (1942).
[10] Collaud, C., to Givaudan, U.S. 2234551 (1942).
[11] Yanada, T., S. Yoshikawa, T. Yasukawa, to Soda Koryo, Japan 41 7770 (1966).
[12] Fumio, T., K. Tsuneo, H. Kazuo, to Hasegawa Koryo Co., Japan 4725071 (1972).
[13] Wilke, G., Angew. Chem. **69**, 397 (1957), **75**, 10 (1963).
[14] Hopp, R., K. Bauer, to Haarmann & Reimer GmbH, DP 2136496 (1973).
[15] Becker, J. J., to Firmenich + Cie., DP 2026056 (1970).
[16] Becker, J. J., G. Ohloff, Helv. Chim. Acta **54**, 2889 (1971).
[17] Mahajan, J. R., G. A. L. Ferreira, H. C. Araujo, J. Chem. Soc. Chem. Comm., **1972**, 1078.
[18] Bauer, K., A. Körber, to Haarmann & Reimer, DAS 2731543 (1977).
[19] Zarkharkin, L. I., A. P. Pryanishnikov, V. V. Guseva, U.S.S.R. 521274 (1974).

13. READILY ACCESSIBLE INTERMEDIATES OF MACROCYCLIC MUSKS

D. J. Hagena* and K. A. Bauer*

Summary β-Methyl substituted α,ω-dicarboxylic acids and their derivatives are excellent starting materials for the synthesis of macrocyclic musks including dilactones and the ketone *d,l*-muscone. Until recently these β-methyl substituted compounds were very difficult to synthesize, and consequently they failed to attain industrial significance. A newly discovered synthetic pathway involves reacting the mono acid chloride mono ester of a dicarboxylic acid (or the bis acid chloride) with 3-methyl but-2-enoate esters to afford unsaturated ketodiesters. The latter are reductively converted in high yield into the corresponding β-methyl substituted dicarboxylic acids or their derivatives.

Scheme 1

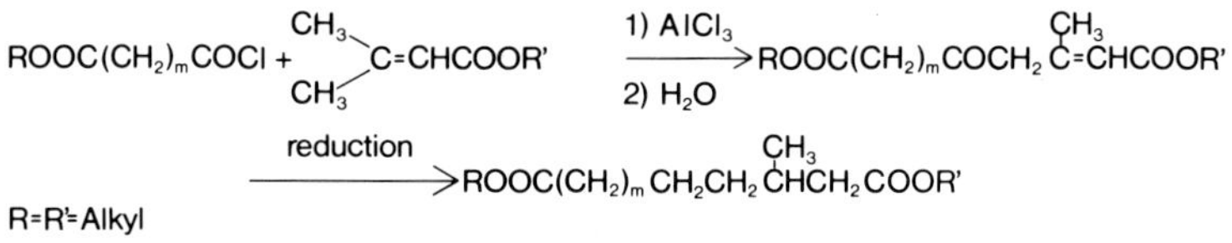

Introduction

The β-methyl substituted carbonyl group is a structural feature frequently found in natural products. Terpene aldehydes, terpene ketones and muscone are typical examples in the area of fragrance substances (Fig. 1). Synthetic procedures which enable the facile insertion of a methyl group at the β-carbon (C3) are consequently of considerable importance. Although there are

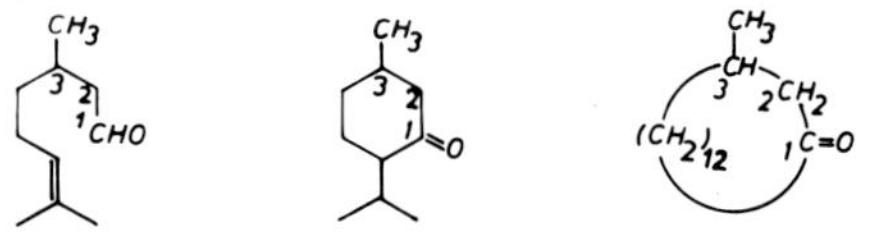

CITRONELLAL MENTHONE MUSCONE

Fig. 1 Some natural products containing a β-methyl substituted carbonyl group.

* D. J. Hagena and K. A. Bauer, Haarmann & Reimer GmbH, 3450 Holzminden, West Germany.

numerous methods available for the introduction of substituents at the α-carbon atom (C2) using mild reaction conditions [1], the synthesis of C3-alkyl substituted oxo compounds via direct electrophilic attack is not possible due to the lack of reactivity of the β-carbon atom (Fig. 2). Technically, such a reaction would be equivalent to a Michael addition reaction with an acceptor of reversed polarity (Fig. 3). During the last few years, much effort has been directed towards the search for such "homoenolate synthons" as they are called [2]. A number of these

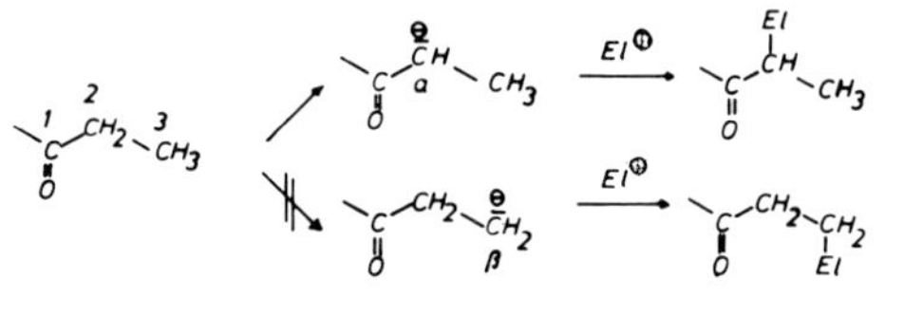

Fig. 2 Comparative reactivity of the α- and β-carbon.

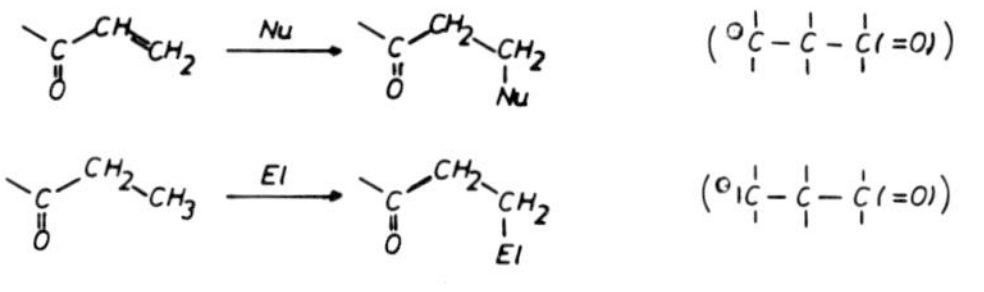

Fig. 3 The Michael addition (upper), and the Michael addition with an acceptor of reversed polarity (lower).

synthons are illustrated in Table 1. These starting materials are by no means cheap as they are often the products of multistep, sometimes very complex reactions. Only the enamines (Table 1), which are synthesized from the ketones and the corresponding secondary amines, can be obtained easily and relatively cheaply, and enamines have proved to be very well suited for the introduction of substituents at the β-carbon atom [3, 8, 9]. However, their application in the introduction of a β-methyl group in macrocyclic ketones or in α,ω-dicarboxylic acids, after suitable derivatization, is hardly feasible due to both technical and economic reasons. Thus, for example, extremely reactive lithium bases (e.g., *t*-butyl lithium) are necessary, and highly reactive anionic intermediates are involved.

Syntheses of 3-Methyl Thapsic Acid Employing β-Methyl Substituted Building Blocks

One way of obtaining carbonyl compounds displaying a β-alkyl substituent is to use starting materials which already have a stable β-alkyl substituent. Unfortunately, such substituted building blocks are quite rare, and usually can only be obtained by a complex series of reactions [12]. Consequently, this approach has remained largely of academic interest with next to no

TABLE 1

SYNTHON		*LITERATURE*
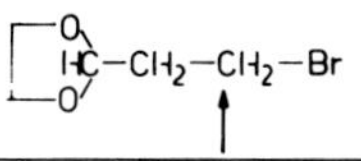	$^{\ominus}ICH_2-CH_2-CHO$	G.BÜCHI, W.WÜEST J.ORG.CHEM., 34, 1122, (1968)
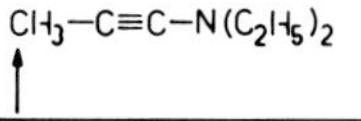	$^{\ominus}ICH_2-CH_2-CHO$	E.J. COREY, D.E. CANE J. ORG.CHEM., 35, 3405 (1970)
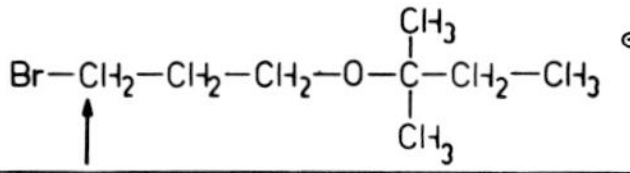	$^{\ominus}ICH_2-CH_2-CH_2OH$	P. EATON, G.F. COOPER R.C. JOHNSON, R.H. MUELLER J. ORG.CHEM., 37, 1947 (1972)
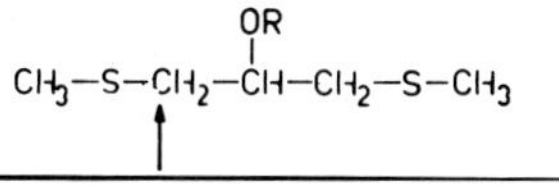	$^{\ominus}ICH=CH-CHO$	E.J. COREY, B.W. ERICKSON R. NOYORI J.AM.CHEM.SOC. 93, 1724 (1971)
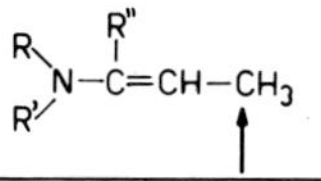	$^{\ominus}ICH_2-CH_2-C(=O)-R''$	H. AHLBRECHT, G. RAUCHSCHWALBE SYNTHESIS, 7, 417 (1973) H. AHLBRECHT, J. EICHLER SYNTHESIS, 9, 672 (1974)
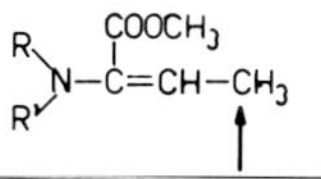	$^{\ominus}ICH_2-CH_2-C(=O)-C(=O)-OCH_3$	H. AHLBRECHT C. VONDERHEID SYNTHESIS, 8, 512 (1975)
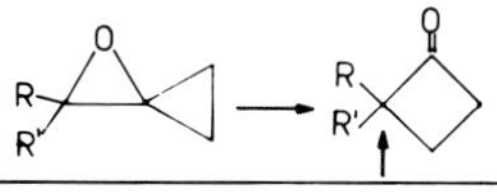	$^{\ominus}ICH_2-CH_2-COOCH_3$	B.M. TROST, T. AMBELANG M.J. BOGDANOWICZ TETRAHEDRON LET. 12, 923 (1973) J.AM.CHEM.SOC. 94, 4777 (1972)
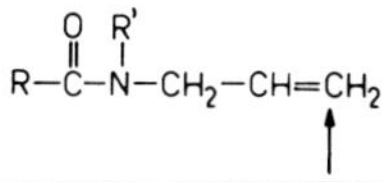	$^{\ominus}ICH_2-CH_2-CHO$	T. HASSEL D. SEEBACH ANGEW. CHEM. 91, 427 (1979)

industrial importance. For example, Figure 4 shows a somewhat modified Hunsdiecker synthesis of 3-methyl thapsic acid [12], which serves as an intermediate in the synthesis of muscone. Understandably, this classical synthesis has not attained any industrial significance due to its multistep nature.

An entirely new concept for the synthesis of 3-methyl thapsic acid and its homologues is one in which the β-methyl substituent is introduced by means of an available preformed C_5 building block (Fig. 5). In contrast to the hitherto known syntheses, this procedure enables the 3-methyl-α,ω-dicarboxylic acid to be obtained in high yields and in only a few reaction steps, providing a ready source of β-substituted precursor for the synthesis of muscone and other macrocyclic musks. Thus, if a mono acid chloride of an α,ω-dicarboxylic acid ester (*1*) is reacted with the ester of 3-methyl but-2-enoic acid (*2*) under Friedel-Crafts reaction conditions, then acylation of one of the methyl groups [13, 14] of (*2*) takes place and the corresponding isomeric unsaturated esters of the ketodicarboxylic acids (*3–5*) are obtained (Fig. 5). If necessary, compounds (*4*) and (*5*) can be stabilized by isomerization to (*3*), which is the main product of the reaction. The isomeric keto diesters (*3–5*) can be reduced either directly to the

3-methyl-α,ω-dicarboxylic acid (*7*) or by a two step reaction via the intermediate saturated compound (*6*) (Fig. 6). Thus, 3-methyl thapsic acid (*7*, n = 10) is the product of this reaction sequence if the diester obtained from the mono ester of 1,12-dodecanedioic acid mono acid chloride is reduced.

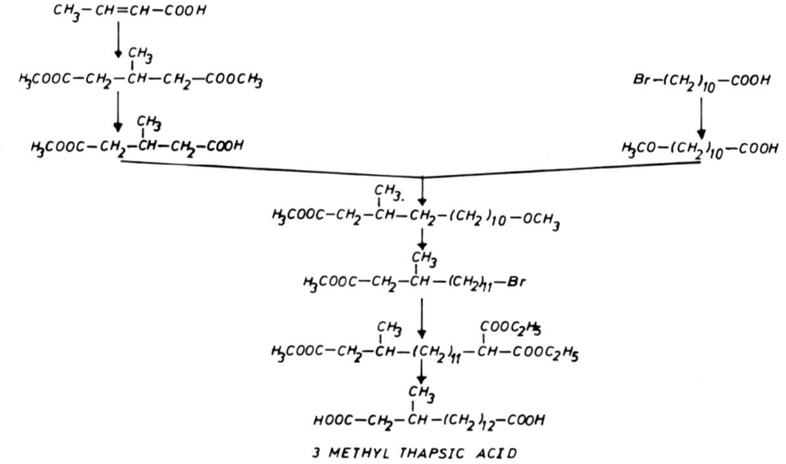

Fig. 4 A multistep synthesis of 3-methyl thapsic acid.

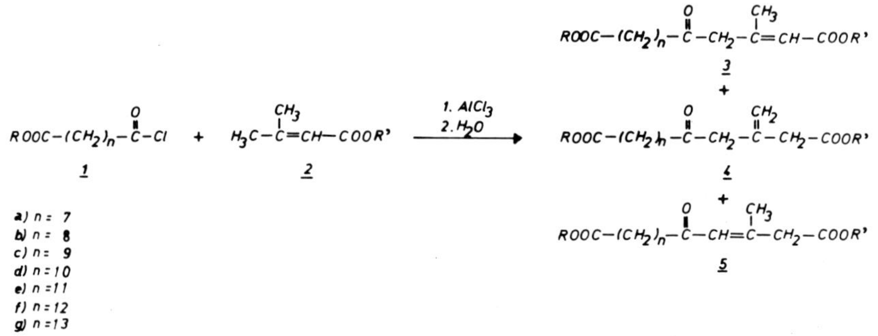

Fig. 5 Condensation of a mono acid chloride of an α,ω-dicarboxylic acid ester (*1*) with the ester of 3-methyl but-2-enoic acid (*2*) under Friedel-Crafts reaction conditions. The R groups are preferentially methyl or ethyl.

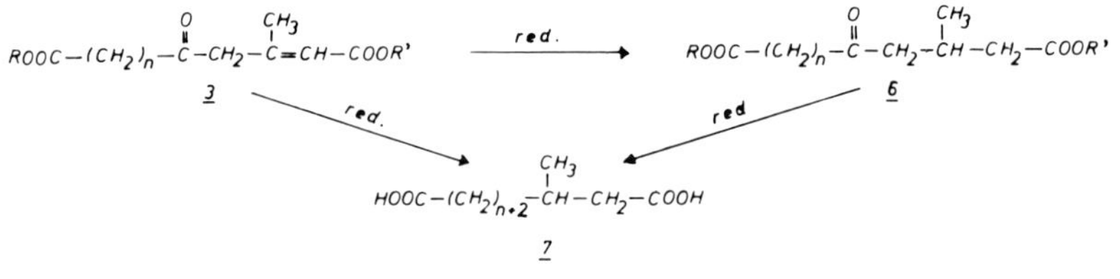

Fig. 6 Reduction of the keto diester [*3*] to the 3-methyl-α,ω-dicarboxylic acid [*7*].

Syntheses of Muscone

Extremely pure, racemic muscone can be obtained upon the cyclization of 3-methyl thapsic acid diester, di-acid chloride [15] or dinitrile [16]. This reaction, in which the carboxyl carbon of the intermediate is lost, is illustrated in Figure 7. Figure 8 shows the 360 MHz NMR spectrum of

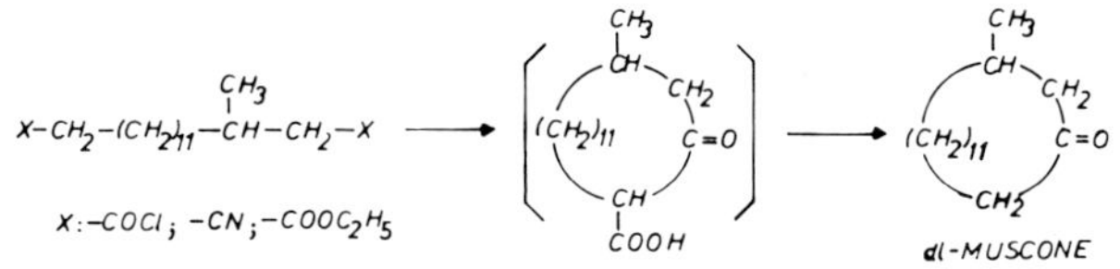

Fig. 7 Cyclization of 3-methyl thapsic acid to *d,l*-muscone.

pure *d,l*-muscone which was obtained from pure 3-methyl thapsic acid synthesized according to the above procedure. The next lower homologue of 3-methyl thapsic acid (3-methyl pentadecandioic acid) can also be obtained in good yields using this synthetic method. Its di-ethyl ester can then be converted, via the acyloin according to the method of Stoll and

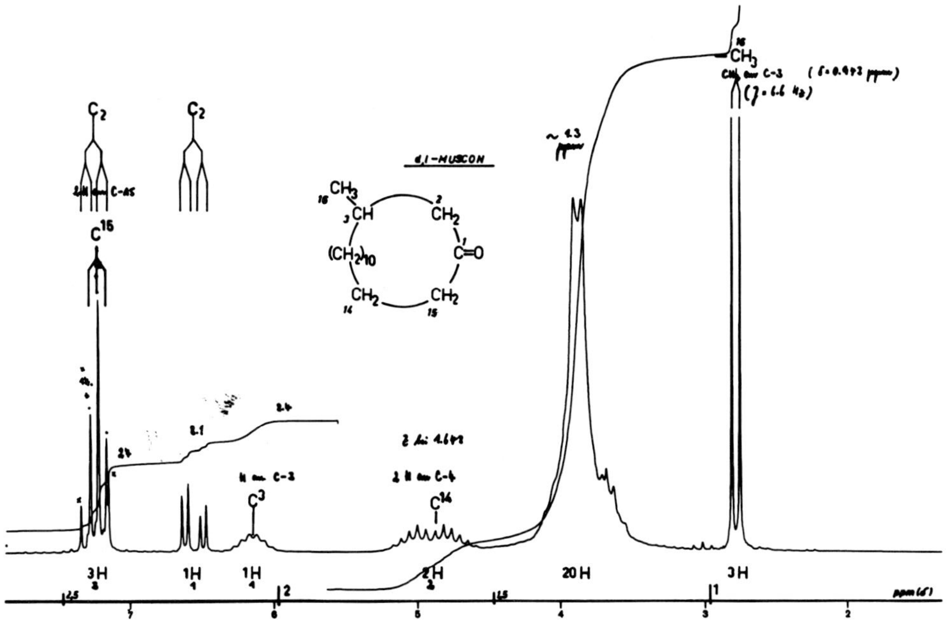

Fig. 8 360 MHz NMR spectrum of *d,l*-muscone.

Commarmont [17], into a mixture of muscone and the isomeric 4-methyl cyclopentadecanone (Fig. 9). This reaction does not involve the loss of a carbon atom.

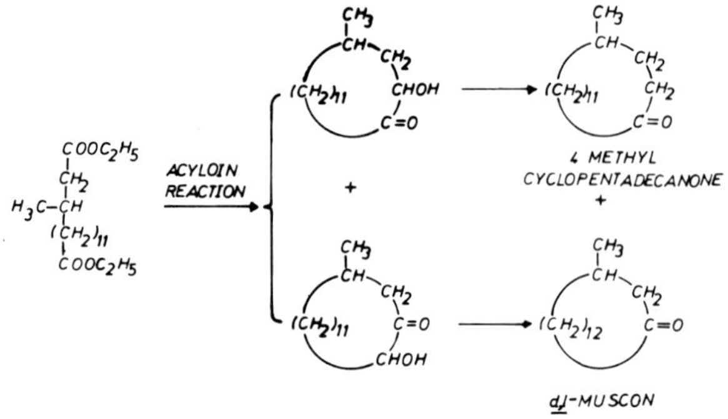

Fig. 9 Conversion of 3-methyl pentadecandioic acid diester to muscone and 4-methyl cyclopentadecanone according to the procedure of Stoll and Commarmont [17].

Apart from the saturated β-methyl substituted dicarboxylic acids, those dicarboxylic acids with additional functional groups can also be used for the synthesis of muscone. Thus, 4-methyl-6-oxopentadecan-1,15-dioic acid can be cyclized via the intermediate acyloin [18], if the 6-oxo group is protected by the formation of a ketal (Fig. 10). Likewise, 3-methyl-5-oxo-alkan-α,ω-dioic acid derivatives can easily be obtained using the same synthetic procedure. These compounds can be cyclized in an analogous manner, thereby opening up entirely new ways for the synthesis of *d,l*-muscone as illustrated in Figure 11.

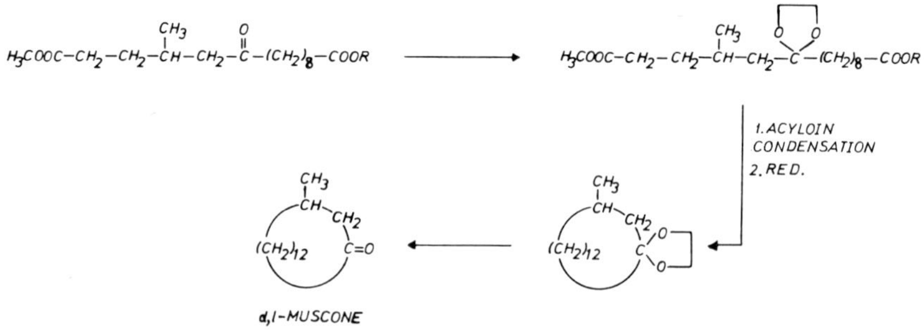

Fig. 10 Cyclization of 4-methyl-6-oxopentadecan-1,15-dioate to *d,l*-muscone.

The reaction scheme described for the synthesis of the 3-methyl-α,ω-dicarboxylic acids (Fig. 5, 6) can also be carried out using the more easily obtainable di-acid chlorides instead of the mono acid chlorides of dicarboxylic acids (Fig. 12). However, the yields of the compounds with structure (*12*) are somewhat lower than those obtained from the *mono* acid chlorides because of a side reaction which leads to the formation of the di-addition product (*10*) (Fig. 12).

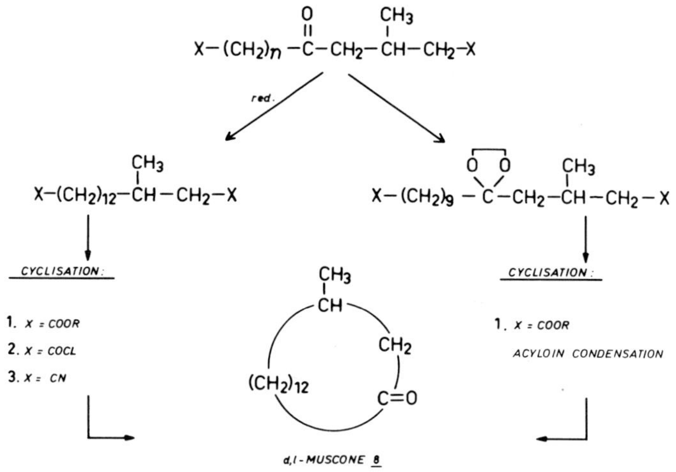

Fig. 11 Alternate routes from 3-methyl-5-oxo-alkan-α,ω-dioic acid derivatives to *d,l*-muscone.

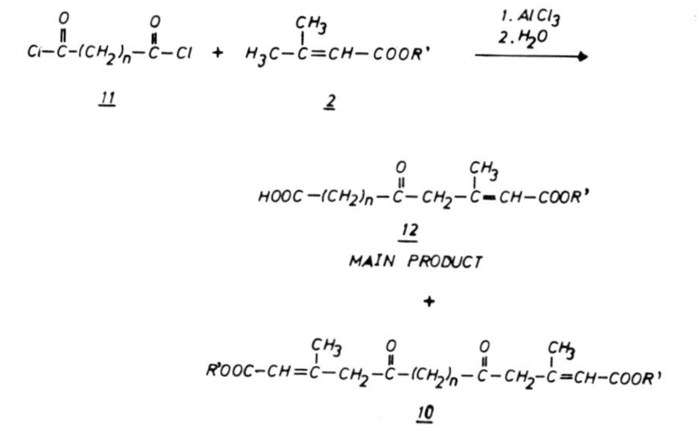

Fig. 12 Synthesis of 3-methyl-α,ω-dicarboxylic acids (*12*) from di-acid chlorides (*11*). For simplicity, double bond isomers of compounds (*10*) and (*12*) are omitted.

Additionally, the unsaturated keto acids (*12*) obtained must be subjected to a more extensive purification procedure before proceeding. The reaction between the di-acid chloride of α,ω-dicarboxylic acids and 3-methyl-but-2-enoic acid can be carried out such that the di-addition compound (*10*) is obtained as the main product. This addition product can then be reductively converted into saturated derivatives of dicarboxylic acids which are suitable intermediates for the synthesis of large ring ketones having two methyl substituents.

Muscone was first isolated from the musk deer by Walbaum in 1906 [19], and it was subsequently shown to be the pure, optically active *R* (–)-stereoisomer [20]. The early syntheses of muscone always yielded racemic product. The first successful, and somewhat complicated, synthesis of the optically pure product was carried out in 1951 by Stalberg-Stenhagen [21] who synthesized both (+) and (–) 3-methyl thapsic acid and subsequently obtained both (+) and (–) muscone via the cyclization of their respective dichlorides. The asymmetrical carbon atom was introduced by using (+)- or (–)-3-methyl glutaric acid as a building block in the synthesis of methyl thapsic acid.

The unsaturated nature of the keto dicarboxylic acid derivatives (*3*) (Fig. 13) led to the assumption that it might be possible to obtain optically active dicarboxylic acids via asymmetrical catalytic reduction using transition metal complexes. Thus, optically active 3-methyl thapsic acid might be available for the synthesis of muscone absolutely *identical* to that found in nature. The preparation of the optically active ester of the ketodicarboxylic acid (*6d*) was achieved by the homogeneous catalytic hydrogenation of the unsaturated ketodiester (*3d*) or its readily obtainable ethylene ketal (*9d*) in the presence of a catalyst with optically active ligands [22]. The optically active 3-methyl thapsic acid (*8d*) was obtained in only 3.6% enantiomeric excess from the ketodicarboxylic acid ester, but in approximately 50% enantiomeric excess from the ketal derivative (Fig. 13). The hydrogenation catalyst (Fig. 14) used was the dimer of chlorobis(ethylene)-rhodium (I) and the chiral ligand was *d*-neomenthyldiphenylphosphine, which can be obtained relatively easily [23].

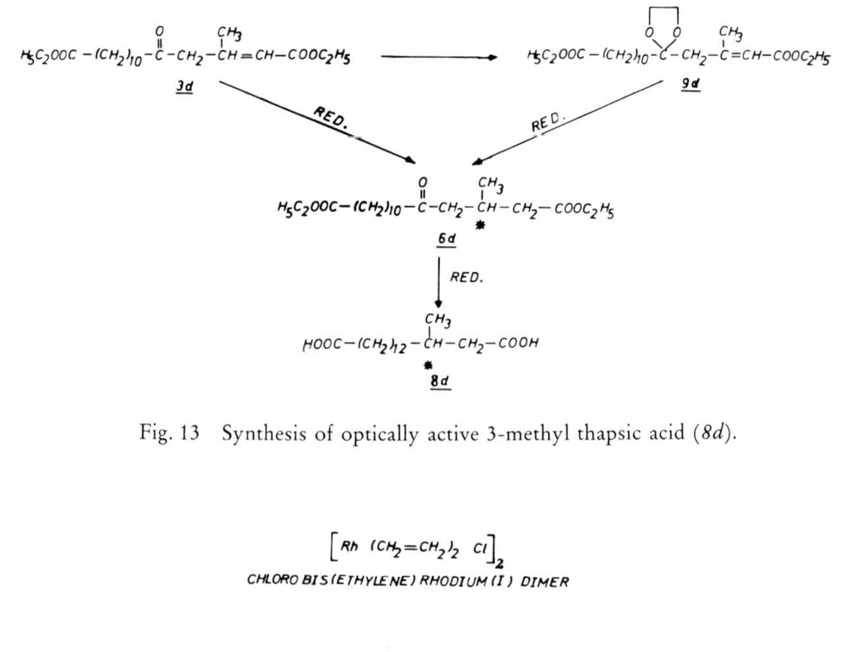

Fig. 13 Synthesis of optically active 3-methyl thapsic acid (*8d*).

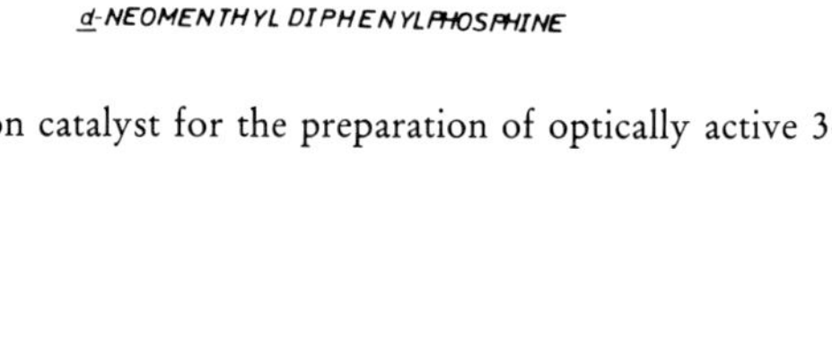

Fig. 14 Hydrogenation catalyst for the preparation of optically active 3-methyl thapsic acid.

Other Applications of the Synthesis

Other compounds that can be synthesized via the intermediates obtained from the reaction between 3-methyl but-2-enoic acid ester and the monoester monoacid chloride of dicarboxylic acids include muscone analogs (Fig. 15, I) and unsaturated large ring ketones with 14- to 16-membered rings (Fig. 15, II). These groups of compounds are described in detail in paper No. 14 of this symposium. The ethylene esters of long chain dicarboxylic acids constitute another well-known class of macrocyclic musks [24, 25]. The 3-methyl dicarboxylic acids synthesized according to the procedure described can be esterified with glycols and thus converted into macrocyclic dilactones (Fig. 15, III). This type of compound, which was hitherto unknown, combines an intense musk odor with a costus root oil note, thereby conveying a considerably more complex odor impression. Additionally, 3-methyl dicarboxylic acid esters can be reduced to their corresponding alcohols which, after conversion to cyclic esters such as macrocyclic carbonates (Fig. 15, IV), also exhibit interesting musk odors [26]. Finally, 3-methyl-ω-hydroxy acids can be obtained, and then used for the synthesis of macrocyclic lactones such as 3-methyl-15-pentadecanolide (Fig. 15, V).

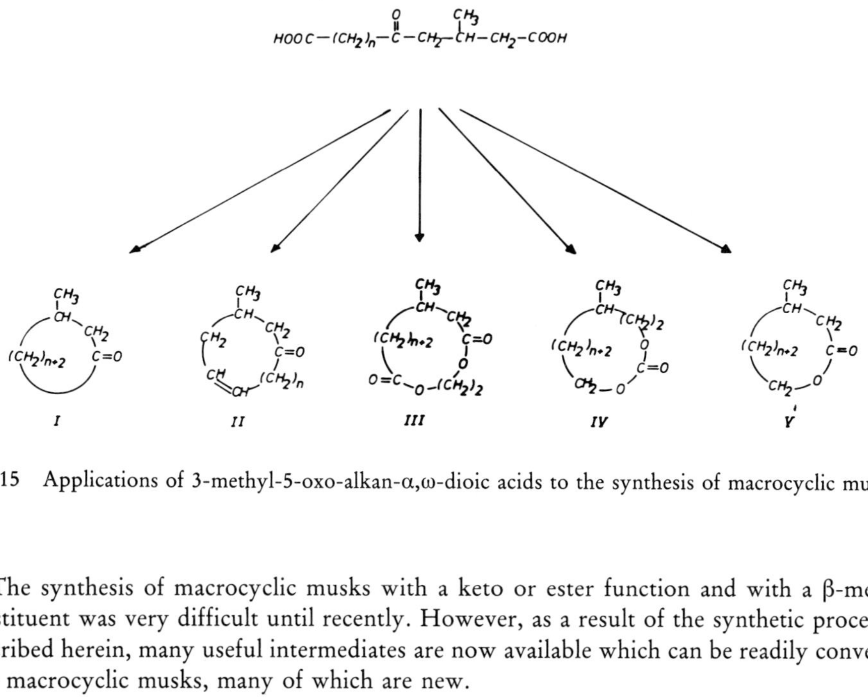

Fig. 15 Applications of 3-methyl-5-oxo-alkan-α,ω-dioic acids to the synthesis of macrocyclic musks.

The synthesis of macrocyclic musks with a keto or ester function and with a β-methyl substituent was very difficult until recently. However, as a result of the synthetic procedure described herein, many useful intermediates are now available which can be readily converted into macrocyclic musks, many of which are new.

References

[1] House, H. O., Modern Synthetic Reactions (second edition), W. A. Benjamin, Inc. Menlo Park, California 1972.
[2] Seebach, D., Angew. Chem. **91**, 259 (1979).
[3] Ahlbrecht, H., G. Rauchschwalbe, Synthesis **1973**, 417.
[4] Büchi, G., W. Wüest, J. Org. Chem. **34**, 1122 (1968).
[5] Corey, E. J., D. E. Cane, J. Org. Chem. **35**, 3405 (1970).
[6] Eaton, P. E., G. F. Cooper, R. C. Johnson, R. H. Mueller, J. Org. Chem. **37**, 1947 (1972).
[7] Corey, E. J., B. W. Erickson, R. Noyori, J. Am. Chem. Soc. **93**, 1724 (1971).
[8] a) Lit. [3]; b) Ahlbrecht, H., J. Eichler, Synthesis **1974**, 672.
[9] Ahlbrecht, H., C. Vonderheid, Synthesis **1975**, 512.
[10] Trost, M., T. Ambelang, M. J. Bogdanowicz, Tetrahedron Lett. **12**, 923 (1973).
[11] Hassel, T., D. Seebach, Angew. Chem. **91**, 427 (1979).
[12] Hunsdiecker, H., Chem. Ber. **10**, 1197 (1942).
[13] Lohaus, G., W. Friedrich, J. P. Jeschke, Chem. Ber. **100**, 658 (1967).
[14] Alkonyi, J., Chem. Ber. **98**, 3099 (1965).
[15] Blomquist, A. T., R. D. Spencer, J. Am. Chem. Soc. **70**, 30 (1948).
[16] Ziegler, K., H. Eberle, H. Ohnlinger, Ann. Chemie **504**, 94 (1933).
[17] Stoll, M., A. Commarmont, Helv. Chim. Acta 31, 1435 (1948).
[18] Mamdapur, V. R., P. P. Pai, K. K. Chakravarti, U. G. Nayak, S. C. Bhattacharyya, Tetrahedron 20, 2601 (1964).
[19] Walbaum, H., J. Prakt. Chem. [2] **73**, 488 (1906).
[20] Klyne, W., J. Buckingham, Atlas of Stereochemistry. Chapman a. Hall London, p. 36 (1974).
[21] Ställberg-Stenhagen, S., Ark. Kemi **3**, 517 (1951).
[22] Valentine, D. Jr., J. W. Scott and literature refs. herein, Synthesis **5**, 329 (1978).
[23] Horner, L., Kontakte (Merck) **3**, 3 (1979).
[24] Spanagel, E. W., H. W. Carothers, J. Am. Chem. Soc. **57**, 929 (1935).
[25] Arctander, S., Perfume and Flavor Chemicals, Vol. I/II Nr. 1214; S. Arctander, Montclair, New York 1969.
[26] Hill, J. W., W. H. Carothers, J. Am. Chem. Soc. **55**, 5039 (1933).

14. INFLUENCE OF MOLECULAR STRUCTURE UPON THE MUSK ODOR CHARACTERISTICS OF MACROCYCLIC KETONES

A. K. Körber* and K. A. Bauer*

Summary A large number of macrocyclic ketones were synthesized and obtained in pure form in order to systematically compare their musk odor characteristics. Properties especially important for perfumers, such as odor, odor threshold, odor intensity and tenacity, were determined for each individual compound, and these properties were then related to the molecular size and the varying structural elements. Some of the new and interesting musks with ketone structures are described in detail and their possible application in perfumery discussed.

Introduction

Comparison of the fragrance qualities of nitro-, polycyclic indane- or tetralin-type musks with the macrocyclic musks always favors the latter. Macrocyclic compounds such as pentadecanolide and muscone elicit a much more natural odor response than their synthetic substitutes. However, the synthetic substitutes do find wide use mainly because of their lower price. In addition to their superior perfumery qualities, the macrocyclic musks are also preferable from the product safety standpoint, in that they are naturally occurring substances of very wide distribution.

For the above reasons, a group of macrocyclic ketones, which are structurally very similar to the naturally occurring cyclopentadecanone, muscone and civetone, were investigated in detail. These ketones are arranged in Figure 1 according to the number of carbon atoms present and the degree of substitution. Included are unsubstituted ketones containing from thirteen to seventeen ring carbons, those with a C2- or C3-methyl substituent, and some unsaturated representatives. According to the literature [1–9], systematic comparisons have been previously performed only with unsubstituted macrocyclic ketones, and occasionally with muscone and civetone. The field of investigation was broadened to include all those ketones in Figure 1, in order to determine those which were most valuable to the perfumer. Included in this study were ketones *13d, 15d, 17d, 17b* and *18c* which appeared to be completely unknown.

* A. K. Körber and K. A. Bauer, Haarmann & Reimer GmbH, 3450 Holzminden, West Germany.

MACROCYCLIC KETONES

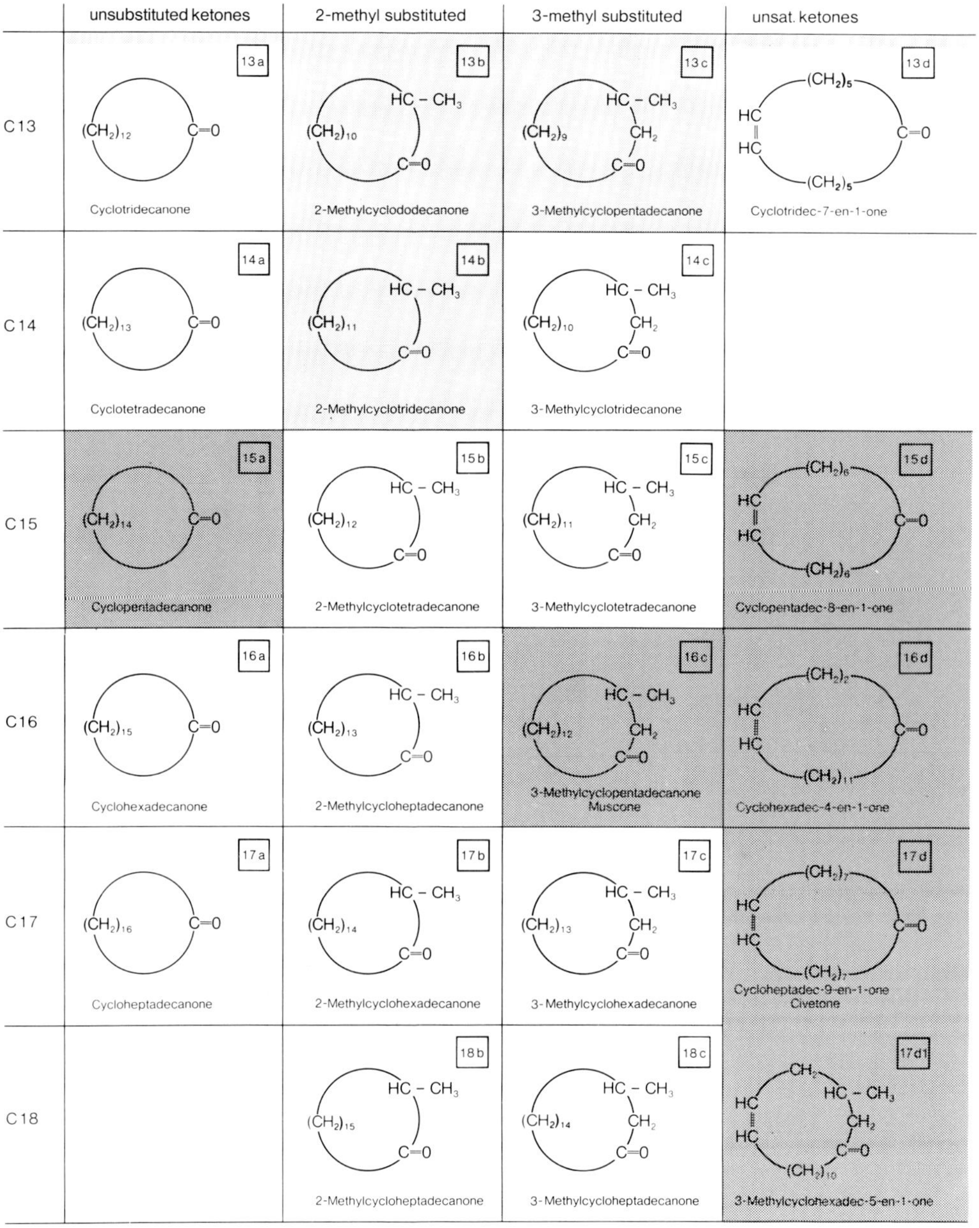

Fig. 1 Macrocyclic ketones.

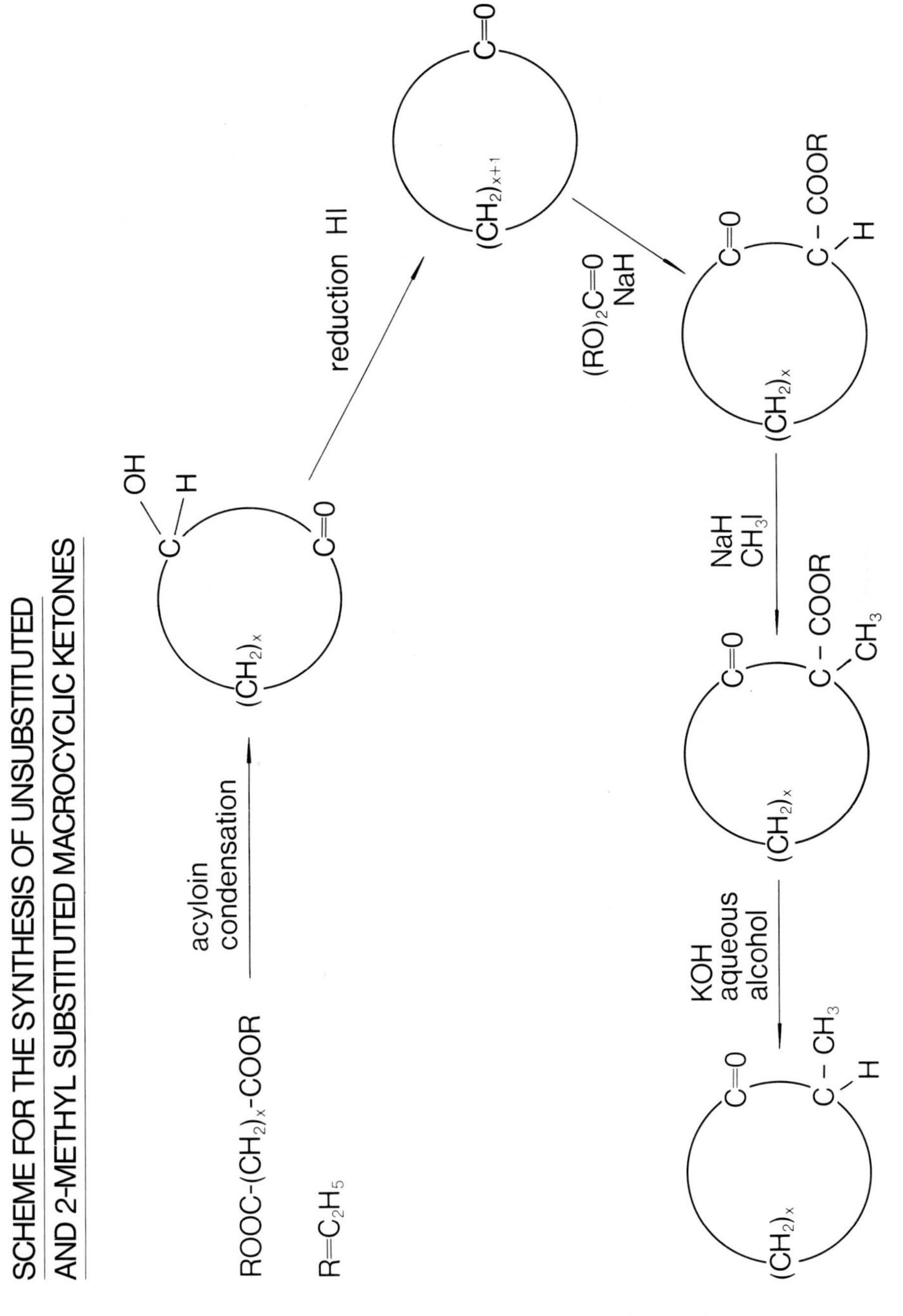

Fig. 2 Scheme for the synthesis of unsubstituted and 2-methyl substituted macrocyclic ketones. $R{=}C_2H_5$.

Syntheses of Macrocyclic Ketones

Figure 2 illustrates the well-known procedure for the synthesis of unsubstituted ketones (i.e., compounds *13a–17a)* via the acyloin condensation of the corresponding dicarboxylic acid esters, followed by reduction of the acyloins with hydriodic acid to the corresponding macrocyclic ketones [10–14]. The 2-methyl ketones were prepared from the unsubstituted ketones via the successive introduction of a carbethoxy group followed by a methyl group, and subsequent removal of the carbethoxy residue [15] (Fig. 2).

The pathways for the synthesis of 3-methyl ketones are illustrated in Figure 3. Path A involves reduction of the methyl-substituted keto diester to form the corresponding diester, ester condensation of the latter while making use of the dilution principle, and subsequent removal of the carbethoxy group to afford the 3-methyl cycloalkanone [16]. The reaction scheme shown in path B involves the protection of the keto group via ketal formation, conversion of the latter into an acyloin and its reduction to a ketal diol. The diol is then converted into the corresponding bromoacetoxy compound upon treatment with hydrogen bromide in glacial acetic acid. Elimination of the bromine followed by decarboxylation yields the corresponding 3-methyl cycloalkenones, which are then subjected to catalytic hydrogenation to yield the 3-methyl cycloalkanones [17, 18]. Civetone, cyclopentadec-8-en-1-one and cyclotridec-7-en-1-one were synthesized from the corresponding unsubstituted keto diesters according to path B. The preparation of 3-methyl keto diesters, from which the 3-methyl ketones are derived, has been previously described (symposium paper No. 13).

Odor Properties

The odor profiles, the odor intensity and its tenacity, and the relative threshold value for each individual ketone were determined in an attempt to characterize and classify these compounds in a meaningful and, as far as possible, objective manner.

Odor Profiles

The odor profiles made use of those characteristic attributes such as fresh, fruity, woody, musk and erogenous-animalic, with the latter two terms usually occurring together. Each attribute was rated using a 1 to 9 point scale and the values obtained are tabulated in Figure 4.

Comparison of the unsubstituted ketones belonging to the C_{12} to C_{17} homologous series shows the first member of the series, cyclododecanone, described using the terms fresh, fruity and woody (Fig. 4). A typical but weak musk odor is first detected for cyclotridecanone, although the fruity, fresh and woody notes typical of the lower homologue still dominate. The ketones *14a* through *17a* exhibit a predominant musk character, while the ketone *17a* has an additional animalic, castoreum-like note.

A comparison of the homologous series comprising the 2-methyl substituted ketones shows that the musk character is first detected in ketone *15b*, and then becomes the main odor component in the compounds *16b–18b*. Generally speaking, the 2-methyl substituted ketones exhibit a musk odor which is considerably weaker than that of the corresponding unsubstituted,

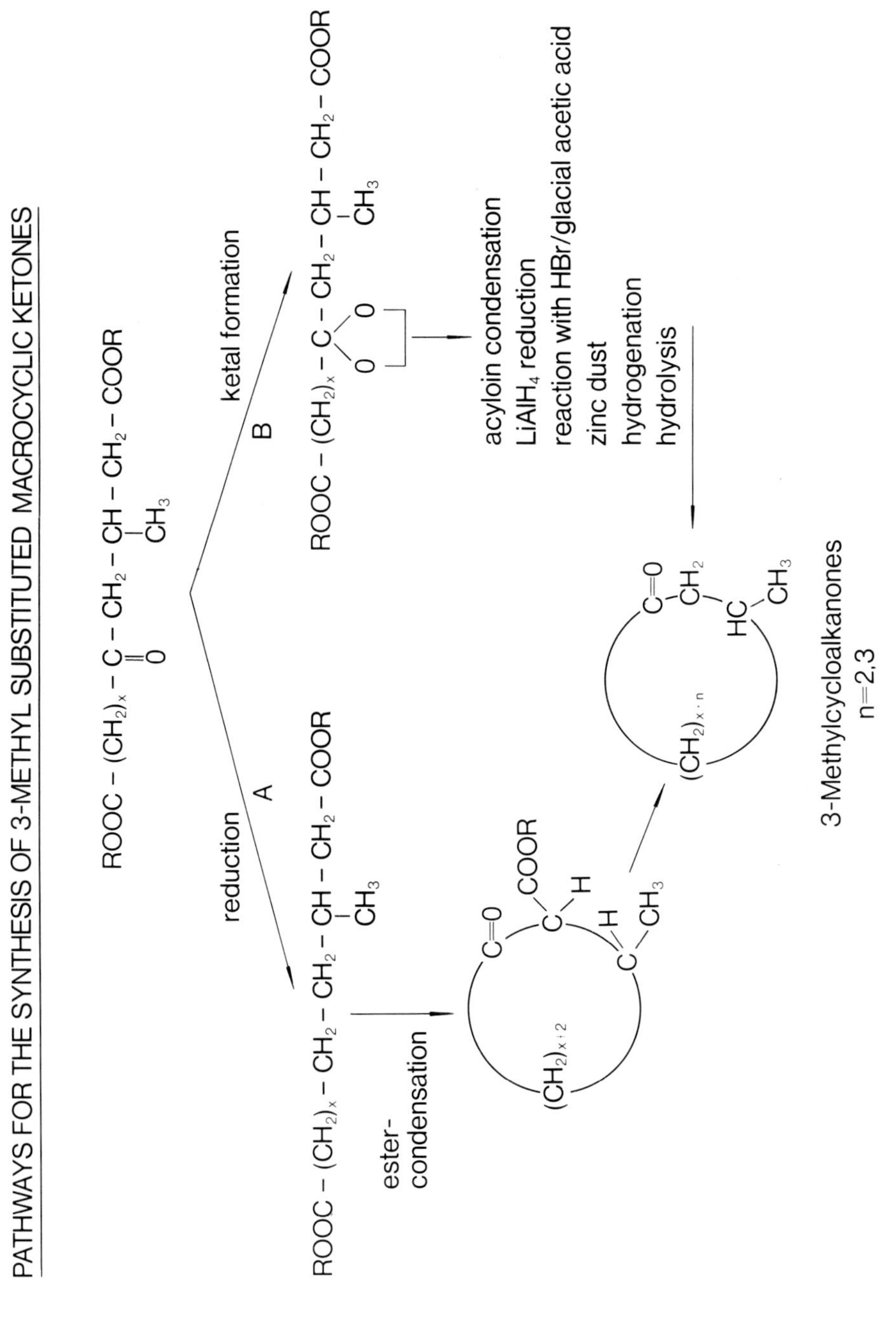

Fig. 3 Pathways for the synthesis of 3-methyl substituted macrocyclic ketones.

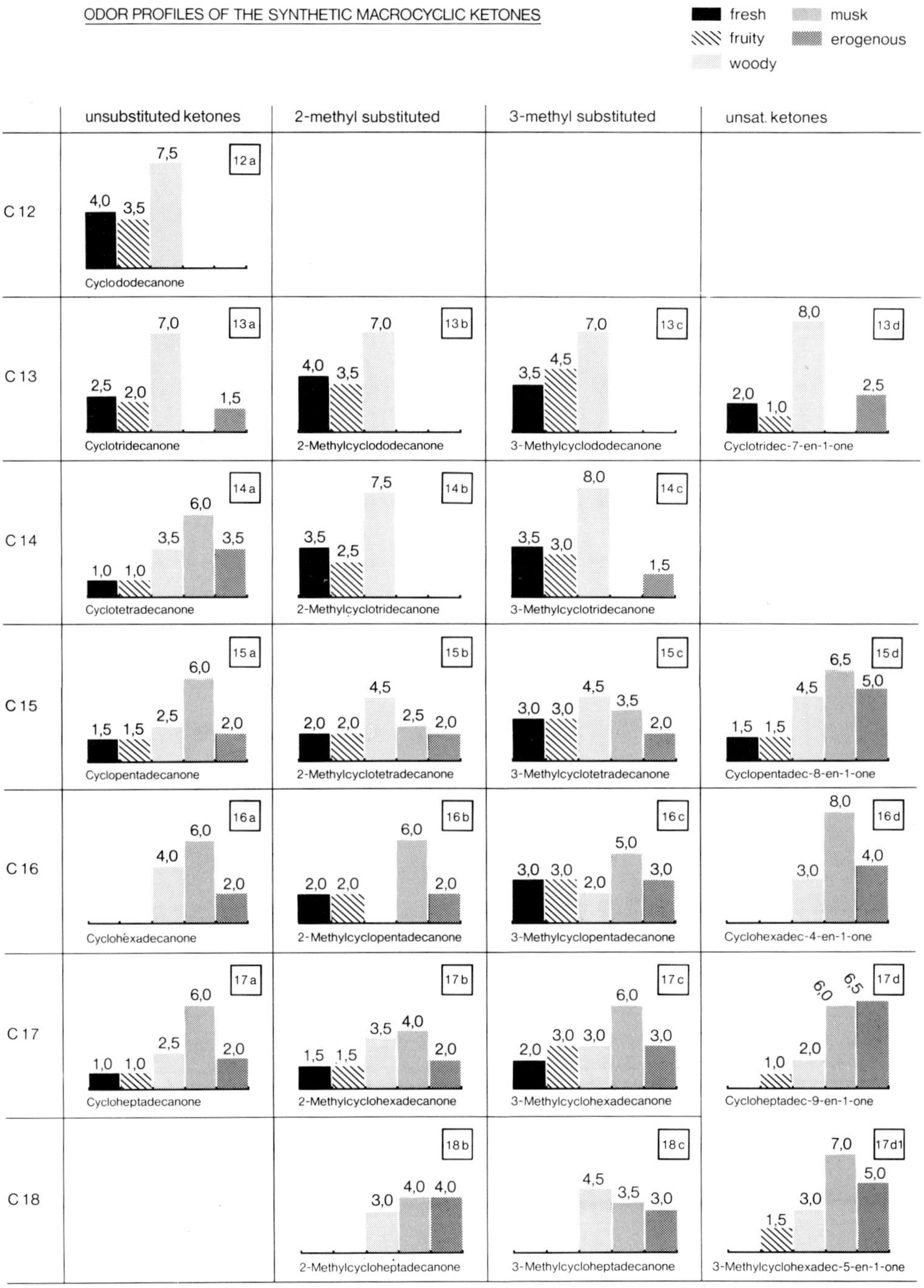

Fig. 4 Odor profiles of the synthetic macrocyclic ketones.

3-methyl substituted, or unsaturated ketones. All of the 2-methyl ketones give a poor and unimaginative overall odor impression.

In the case of the 3-methyl series, the musk character is first detectable in the 13-membered ring ketone. 3-methyl cyclotridecanone (compound *14c*) has a weak animalic component in addition to a strong woody note. The musk note starts to dominate the overall odor impression from 3-methyl cyclotetradecanone *(15c)* onward, and the most attractive members of this series are muscone *(16c)* and homomuscone *(17c)*. The complexity of the odor shown by the 3-methyl-substituted ketones is especially striking. The C_{15}, C_{16} and C_{17} ketones belonging to series *c* have, in addition to the fresh and woody character, a distinct fruity note which balances and generally improves the overall musk impression. This is in contrast to the unsubstituted ketones *15a*, *16a* and *17a*, in which the musk note dominates while the fresh and fruity character is of little importance.

The macrocycles belonging to series *d* are most similar to the unsubstituted ketones as far as their odor characteristics are concerned. They also have a simpler overall fragrance impression, which enables them to be classified more easily. The unsaturated ketones first display a distinct musk character at the 13-membered ring stage. The musk character becomes dominant from compound *15d* onwards, and the woody, fruity and fresh notes diminish in intensity. A conspicuous characteristic of the unsaturated ketones is their very distinct animalic, erogenous note, which is considerably more pronounced here than in the case of the *a*, *b* and *c* series. Cyclotridec-7-en-1-one *(13d)* and cyclopentadec-8-en-1-one *(15d)*, which were included in this survey, were hitherto unknown. Both compounds are symmetrical and, thus, are lower homologues of civetone *(17d)*. The musk ketone, 3-methyl cyclohexadec-5-en-1-one *($17d_1$)*, which was also unknown prior to this study, exhibits fragrance properties which lie between those of homomuscone *(17c)* and civetone *(17d)* (for further discussion of this ketone see symposium paper No. 11).

A line drawn at the boundary where the transition from the fruity, fresh and woody odors to the typical musk note occurs is indicated in Figure 4. This line starts at *13a*, continues via *15b* and *14c*, and runs between *12d* and *13d* since the C_{12}-cycloalkenones show almost no musk character. Based on this comparative study, those ketones which embody the most intense musk and erogenous notes are cyclopentadecanone *(15a)* and cyclohexadecanone *(16a)* in the *a* series, compounds *15c*, *16c* and *17c* in the *c* series, and compounds *15d*, *16d*, *17d* and *$17d_1$* in the *d* series. It is quite striking that it is the unsaturated ketones, with their intense animalic note, which display the typical, strong musk character.

Odor Intensities and Tenacities

Investigation of relative odor intensities, which are summarized in Figure 5, furnished valuable information concerning the perfumistic classification of the macrocyclic ketones. Two percent alcoholic solutions of the ketones were used and were checked over a period of 10 days. It had been previously shown that all the musks underwent a distinct decrease in odor intensity within the first 10 days, and the testing was terminated at this point. The musk ketones were smelled blind each day, were sorted according to their strengths, and then allocated grades from 6 for the most intense to 1 for the weakest sample.

Of all those substances tested, the unsaturated macrocylic ketones have the most intense odors, and this remains so during the entire testing period. Apart from compound *13d*, which

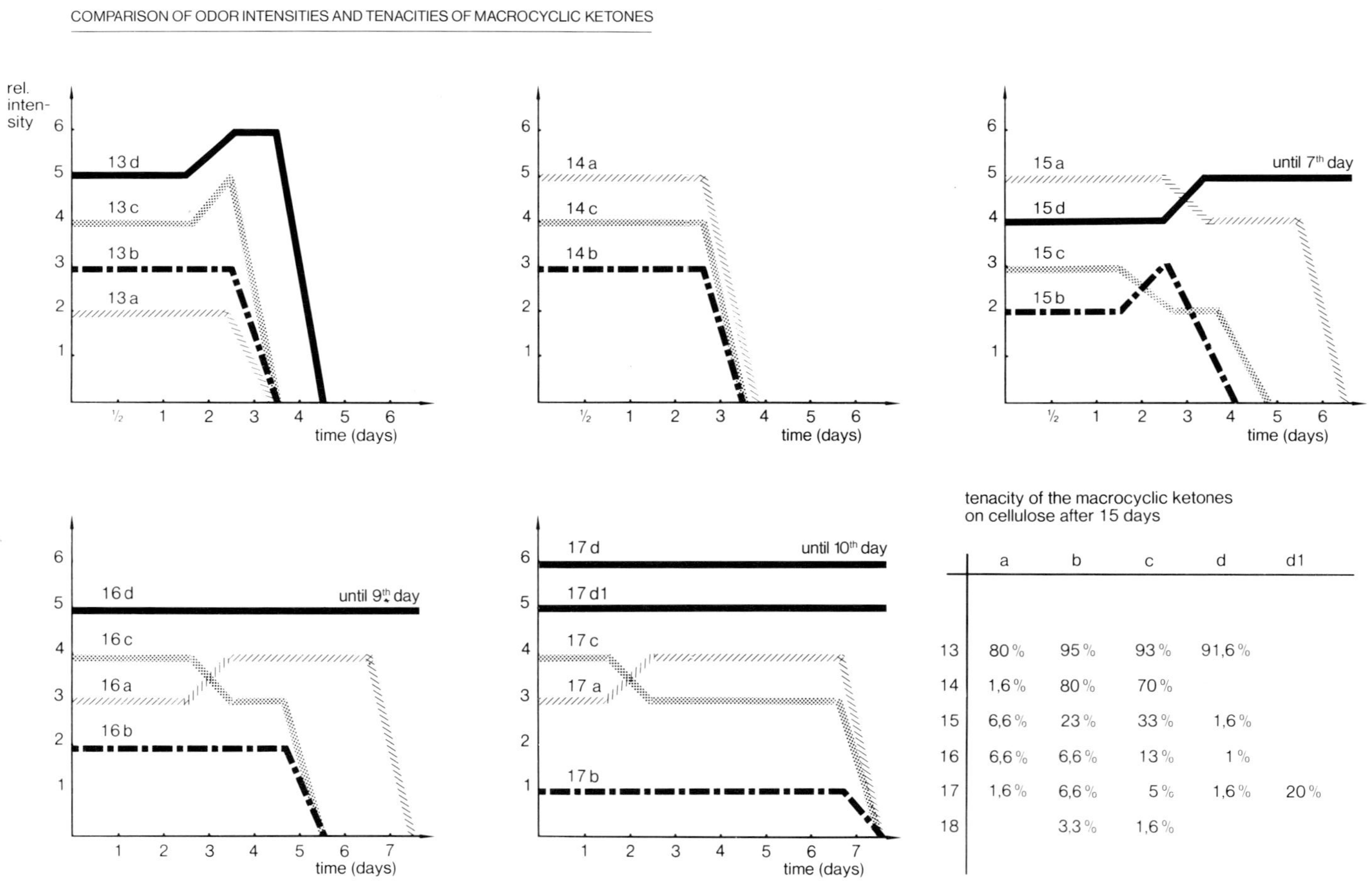

tenacity of the macrocyclic ketones on cellulose after 15 days

	a	b	c	d	d1
13	80 %	95 %	93 %	91,6 %	
14	1,6 %	80 %	70 %		
15	6,6 %	23 %	33 %	1,6 %	
16	6,6 %	6,6 %	13 %	1 %	
17	1,6 %	6,6 %	5 %	1,6 %	20 %
18		3,3 %	1,6 %		

Fig. 5 Comparison of odor intensities and tenacities of macrocyclic ketones. Tenacity values are reported in percent loss.

had completely evaporated after 4 days, all of the compounds in the *d* series are considerably more intense than the corresponding isomers belonging to the *a*, *b* and *c* series. One exception seems to be *15d*, which is initially a little weaker than *15a*, but which exhibits a reversal after 4 days due to its greater tenacity. The 2-methyl substituted cycloalkanones (*b* series) have the weakest odor intensities of all the ketones tested. It should be noted that a similar overall impression was obtained in classifying the odor profiles. It was also found that the intensities of the unsubstituted (*a* series) and 3-methyl substituted (*c* series) ketones lie between those of the 2-methyl substituted (*b* series) and the unsaturated (*d* series) ketones.

The determination of the relative odor intensities showed that the 3-methyl substituted ketones initially have a very intense odor, as do the unsubstituted ketones, but that their odor intensities decrease rapidly between the second and the seventh day depending upon the ring size. These findings paralleled the tenacities (Fig. 5), which were determined using cellulose strips [19]. As expected, the tenacity values (percent evaporated) for ketones of the *a*, *b* and *c* series decrease with an increasing number of carbon atoms in the molecule. Surprisingly, compounds belonging to the *d* series show a high tenacity, and, even after the addition of a stabilizer, less than 2 % of the ketones *15d*, *16d* and *17d* had evaporated after 15 days.

Relative Odor Threshold Values

The relative odor threshold values were determined using dilute solutions of the macrocyclic ketones. Since only the relative threshold values of the macrocyclic ketones were needed, the method of Stevens [20, 21] and the test procedure of Theimer and Davies [22] were modified to suit this requirement. A preliminary test was performed in order to determine which concentration range should be used for these closely related compounds. The ketones were tested in dipropylene glycol, and the concentration range chosen was from 1000 to 0.1 ppm. The test panel consisted of 20 judges, and the test itself was carried out in an air conditioned room with a relative humidity of 50 % at 20 °C. The panel was presented with five coded cellulose smelling blotters comprising a concentration series of 0.1, 1.0, 10, 100 and 1000 ppm. A smelling blotter with pure dipropylene glycol served as a control. The judges were instructed to arrange the blotters according to concentration and to indicate where the musk odor was just detectable.

A standard curve, illustrating the dependency of the degree of correct rating (in percent) as a function of the logarithm of the concentration, was established for each ketone, and from these curves the approximate threshold concentration levels can be obtained. The characteristic curves found for compounds *17a* through *17d* are assembled in Figure 6. For compound *17a* at the 1000 ppm level, all of the judges chose the correct grading, whereas this value decreased to 35 % at the 100 ppm concentration level. Only 7 % of the judges were able to determine the correct order at the next lowest concentration level of 10 ppm. All the macrocyclic ketones were tested in this way, and the data compiled in a series of curves from which several generalizations can be made. The best response curves are obtained for the unsaturated ketones (*d* series). The 3-methyl substituted (*c* series) and the unsubstituted (*a* series) ketones produce very similar curve shapes, although the degree of discrimination is less. The judges exhibited the least discrimination in the case of the 2-methyl substituted (*b* series) ketones; typified by the lowest position for compound *17b* in Figure 6. Further illustrative of this point is the fact that, even at the 1000 ppm level, only 30 % of the panel members were able to detect and correctly classify compounds of

RANGES OF ODOR THRESHOLDS OF THE MACROCYCLIC KETONES

dependency of relative probability
in % upon concentration

conz. ppm	17a	17b	17c	17d	17d1
1000	100	35	100	100	92,8
100	35	7	42,8	50	39,3
10	7	14,2	14,2	42,8	25
1	7	0	0	0	0
0,1	0	0	0	0	0

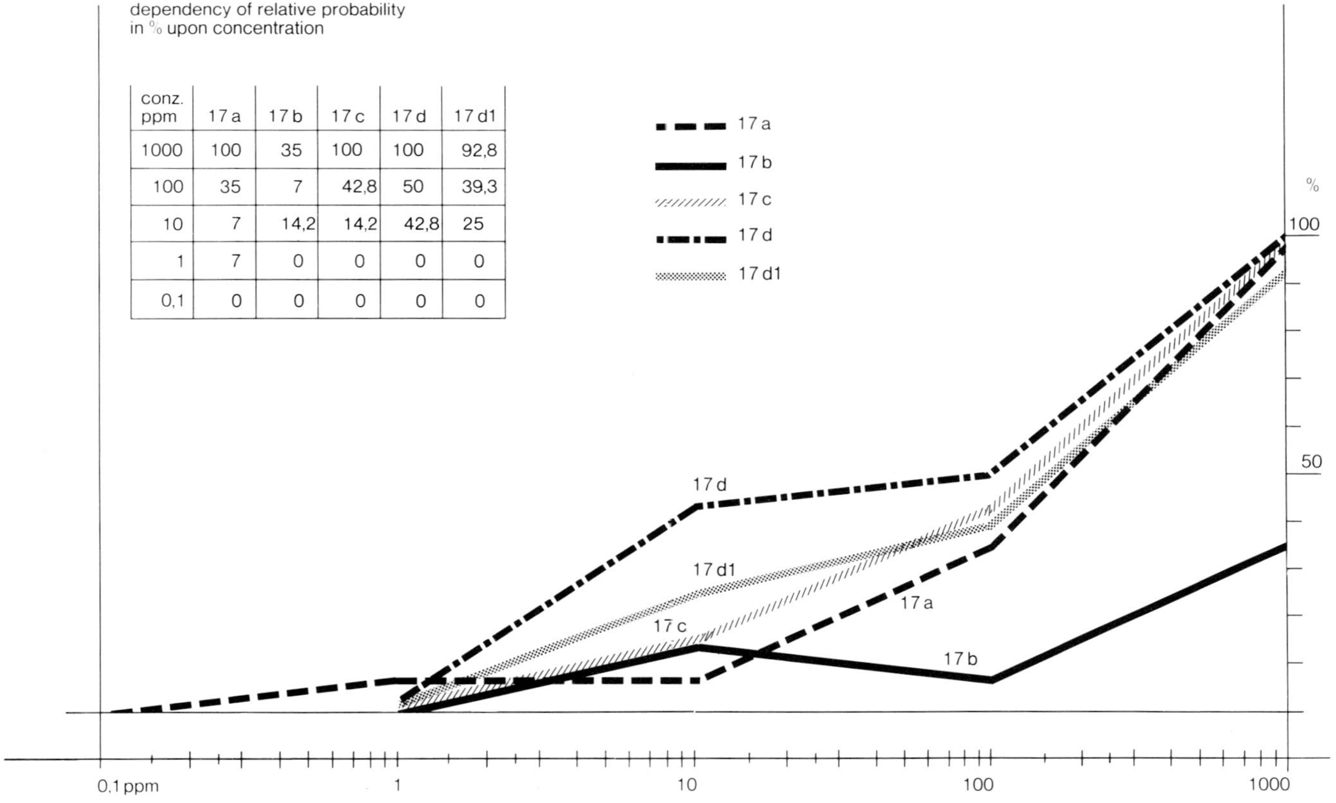

Fig. 6 Ranges of odor thresholds of the macrocyclic ketones.

the *b* series. In general, the tabulated curves indicated that compounds *17d*, *17d*$_1$, *16d*, *15d*, *16c* and *15a* of Figure 1 display low threshold values in the region of 100 ppm.

Conclusions

The combined findings of the odor profile, odor intensity, tenacity and odor threshold studies with the macrocyclic keto musks allow the development of several general conclusions. These conclusions should be borne in mind in developing an optimal strategy for the synthesis of macrocyclic musks.

1. The optimum total number of carbon atoms is 16 or 17.
2. Civetone (*17d*), while somewhat weaker than cyclopentadecanone, is, because of its tenacity,
3. An unsubstituted macrocycle has a relatively clear odor profile in which the musk note dominates and the erogenous note is less noticeable. In this respect the unsubstituted compounds are similar to the 3-methyl substituted compounds, but they do not have the tenacity of the unsaturated macrocyclic ketones.
4. The 2-methyl substituted ketones are the least interesting of those ketones investigated.
5. The 3-methyl substituted ketones have a complex odor profile in which the typical musk and erogenous notes are about as important as the woody, fruity and fresh notes. The fruity note is especially noticeable in this series.
6. A double bond within the ring has an unexpectedly positive effect upon tenacity and odor strength. All the unsaturated macrocycles have the lowest threshold values of those compounds investigated, and the animalic, erogenous notes dominate their odor profiles.

Of all the ketones tested, the following are the most interesting in terms of perfumery use.

1. Cyclopentadecanone (*15a*), because of its very intense fragrance and its clear and uniform musk note, which is only found in the unsubstituted ketones.
2. Civetone (*17d*), while somewhat weaker than cyclopentadecanone, is, because of its tenacity, ideally suited for combination with relatively volatile musks such as cyclopentadecanone or cyclopentadecanolide. Furthermore, civetone can be used to achieve an erogenous, animalic note.
3. Muscone (*16c*), because of its very intense and complex odor profile, which shows fresh and fruity nuances.
4. Cyclohexadec-4-en-1-one (*16d*), due to its intense fragrance, very good tenacity values, and its excellent performance in the threshold determination test.
5. Cyclopentadec-8-en-1-one (*15d*), a hitherto unknown macrocyclic ketone, which exhibits a strong and natural musk note with a distinct erogenous component. The combination of an intense odor with a high degree of tenacity provides a more or less constant musk note.
6. 3-methylcyclohexadec-5-en-1-one (*17d*$_1$), another previously unknown ketone, which embodies excellent tenacity, odor strength and threshold values typical of the macrocycles belonging to the *d* series. Combined with these properties is a complex odor profile, due directly to the 3-methyl substituent, and typical of muscone or homomuscone.

Acknowledgement

We thank K. J. Bork and G. Mansfeld who were involved in the evaluation of the macrocyclic ketones.

References

[1] Ruzicka, L., M. Stoll, H. Schinz, Helv. Chim. Acta **9**, 230, 249, 716, 1008 (1926).
[2] Hill, J. W., W. H. Carothers, J. Am. Chem. Soc. **55**, 5039 (1933).
[3] Blomquist, A. T., J. Prager, J. Wolinsky, J. Am. Chem. Soc. **77**, 1804 (1955).
[4] Blomquist, A. T., J. Wolinsky, J. Am. Chem. Soc. **77**, 5423 (1955).
[5] Ruzicka, L., M. Pfeiffer, Helv. Chim. Acta **11**, 686 (1928).
[6] Ruzicka, L., Helv. Chim. Acta **17**, 1308 (1934).
[7] Stoll, M., Seifen **46**, 136 (1939).
[8] Ruzicka, L., W. Scherrer, C. F. Seidel, Helv. Chim. Acta **15**, 1459 (1932).
[9] Beets, M. G. J., Structure-Activity Relationship in Human Chemoreception, Applied Science Publishers Ltd., London (1970).
[10] Prelog, V., L. Frenkiel, M. Kobelt, P. Barmann, Helv. Chim. Acta **30**, 1815 (1947).
[11] Stoll, M., J. Hulstkamp, Helv. Chim. Acta **30**, 1822 (1947).
[12] Hansley, V. L., J. Am. Chem. Soc. **57**, 2303 (1935), U.S. Pat. 2228268 (1941).
[13] Stoll, M., Helv. Chim. Acta **30**, 1834, 1839 (1947).
[14] Stoll, M., Helv. Chim. Acta **31**, 1082 (1948).
[15] Rhoads, S. J., J. C. Gilbert, A. W. Decora, T. R. Garland, R. J. Spangler, M. J. Urbigkit, Tetrahedron **19**, 1625 (1963).
[16] Leonard, N. J., C. W. Schimelpfennig, jr., J. Org. Chem. **23**, 1708 (1958).
[17] Nair, M. S. R., H. H. Mathur, S. C. Bhattacharyya, J. Chem. Soc. **1964**, 4154.
[18] Chakravarti, K. K., U. G. Nayak, S. C. Bhattacharyya, V. K. Balakrishnan, R. K. Razelan, J. Chem. Soc. **1965**, 3475.
[19] Sturm, W., G. Mansfeld, Chemiker Zeitung **99**, 69 (1975).
[20] Stevens, K. L., D. G. Guadagni, D. J. Stern, J. Sci. Food. Agr. **21**, 590 (1970).
[21] Haring, H. G., F. Rijkens, H. Boelens, A. van der Gen, J. Agr. Food Chem. **20**, 1018 (1972).
[22] Theimer, E. T., J. T. Davies, J. Agr. Food Chem. **15**, 6 (1967).

15. CUTANEOUS TOXICITY AND THE SENSITIZATION PROPERTIES OF CHEMICALLY DEFINED FRAGRANCE SUBSTANCES: REVIEW AND CHEMICAL STUDIES OF THE MECHANISM OF SENSITIZATION

C. Benezra*

Summary Many chemical compounds are irritants or sensitizers. Such irritant properties are often concentration dependent, and can be studied with animal models. According to the classical definition of Gell and Coombs, allergies can be divided into four types, and it is the forth type, delayed hypersensitivity, that is of interest here. Recent research has helped to clarify certain aspects of the cellular mechanism of delayed hypersensitivity. Haptens are small molecules which have to bind macromolecules (i.e., proteins) to become complete antigens. A hapten can be considered as an electrophile and a carrier protein as a nucleophile, and formation of a complete allergen (or antigen) implies the formation of a covalent bond. Thus, to some extent, the immunological activity of certain compounds can be predicted. Some known and potential haptens (incomplete allergens) are aldehydes, ketones, lactones, phenols, unsaturated compounds and few esters. A structure-activity relationship allows an understanding of the mechanism of allergic contact dermatitis, and therefore a prediction of hazards due to contact with chemicals.

Introduction

Ideally, when dealing with a new product the prediction of its action with the skin could avoid a number of problems; yet, is prediction possible? While this important topic will be considered later, it is first necessary to review the main adverse effects of chemical substances on the skin. These are irritation and photoirritation (phototoxicity), allergic contact dermatitis (ACD) and photoallergy, leucodermia (depigmentation of the skin), carcinogenesis, contact urticaria, and general toxicity (by skin penetration). Of all these adverse cutaneous effects, irritation and allergy are the most frequently encountered. Leucodermia will be discussed only in passing.

Another preliminary question that must be addressed is, what is skin? In each cm^2 of skin there are 3 million cells, 15 sebaceous glands, 1 m of blood capillaries, 4 m of nerves and 300

* Claude Benezra, Laboratoire de Dermato-Chimie, Université Louis Pasteur de Strasbourg, Clinique Dermatologique, CHR de Strasbourg, 67091 Strasbourg, France.

sensory cells. A cosmetic advertisement states "Madam, your skin is your most beautiful dress, know how to keep it beautiful." Skin is the first barrier against aggressors which, for the most part, are irritants and allergens.

Irritation and Photoirritation (Phototoxicity)

Anyone exposed to an irritant is, in principle, able to develop a cutaneous reaction ranging from simple redness to necrosis. The intensity of the reaction will depend, of course, on the individual, the time of exposure, and the concentration of the irritant [1, 2]. The next obvious question then is whether the irritancy of a compound can be predicted by simple examination of the chemical structure. Unfortunately, the answer is no. There are so many kinds of irritants (acids, bases, organic solvents, detergents, etc.) that any classification based on chemical families is useless. The only way to predict the irritant properties of a given chemical is to use an animal model, and most often the rabbit is the model of choice [3, 4]. Compared to man, the rabbit tends to overreact to compounds; however, this model is good for strong irritants or non-irritants. For moderate irritants, the guinea-pig is preferred. The latter animal is also an excellent model for studying allergic contact dermatitis (ACD).

Although irritants cannot be easily classified, photoirritants (i.e., irritants which are active only in the presence of light) are generally extensively conjugated cyclic molecules with some fluorescent properties [5]. Furocoumarins (psoralens) are important photoirritants (Fig. 1), as are phenothiazinic derivatives and several other organic compounds (Fig. 2). Bergapten (Fig. 1),

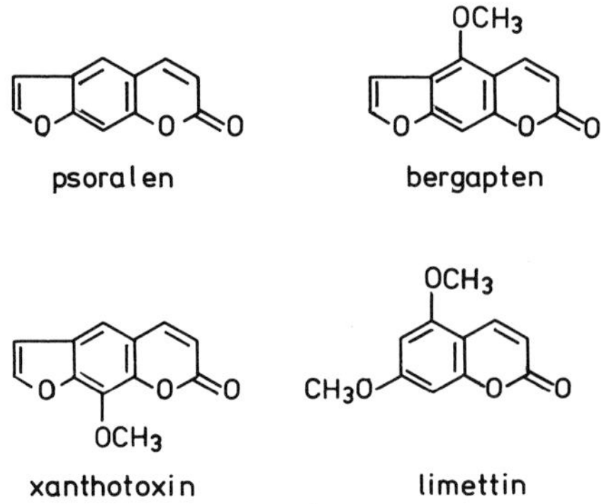

Fig. 1 Furocoumarin photoirritants (Limettin is not toxic).

a component of bergamot oil, is responsible for "berlock dermatitis", and bergamot oil should contain less than 0.001 % bergapten in order to be inoccuous [6]. Another furocoumarin, xanthotoxin (8-methoxypsoralen or 8-MOP), used for the treatment of skin diseases such as vitiligo and psoriasis, is also phototoxic. Another component of bergamot oil, limettin, is not toxic. The irritant properties of psoralens are due to their linking to the pyrimidine bases of DNA, forming four-membered rings [7] (Fig. 3).

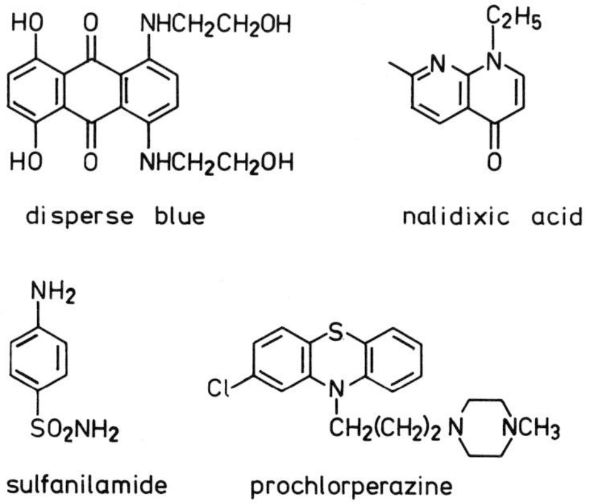

Fig. 2 Other photoirritants.

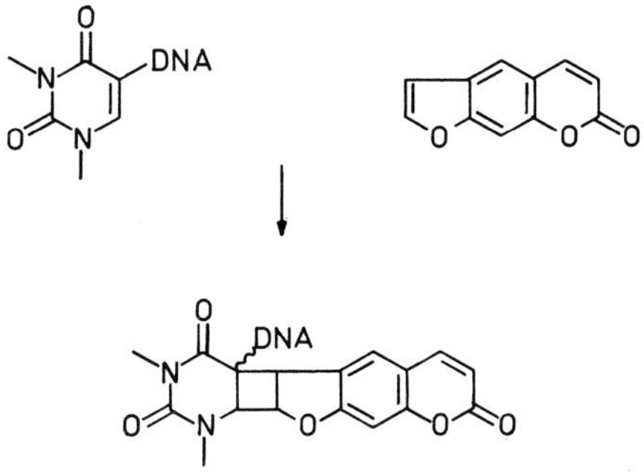

Fig. 3 Irritant properties of psoralens result from their linkage to the pyrimidine bases of DNA.

Allergic Contact Dermatitis

Although any individual is, in principle, sensitive to irritants and photoirritants, not everyone is allergic to allergens. According to the classical definition of Gell and Coombs [8], there are four main types of allergies: type I, immediate or anaphylaxis (hay fever, asthma, etc.); type II, cytotoxicity (incompatabilities of blood groups); type III, semidelayed or Arthus type; and type IV, delayed hypersensitivity (allergic contact dermatitis, homograph rejections). The first three types are characterized by the presence of circulating antibodies (immunoglobulins) in the blood of the patient, while in the fourth type, which is the concern of this paper, these circulating antibodies are absent.

Mechanism of Allergic Contact Dermatitis

Some products produce a delayed reaction, an eczema, when brought into contact with the skin. These products, which are low molecular weight substances, are sensitizers or haptens (i.e., incomplete antigens or allergens). In order to become true allergens (complete allergens), they must bind to a carrier protein. The cellular mechanism of ACD has been reviewed recently [9], and a schematic outline of the series of events follows. The hapten penetrates the skin and binds to a carrier protein. These conjugates (the allergens) then come into contact with T-lymphocytes in the skin (peripheral sensitization) or in the nodes (central sensitization). These T-lymphocytes are now "sensitized", that is they are covered with specific receptors, and in the nodes they differentiate into memory cells, effector cells and suppressor cells. These events occur on the very first contact with the hapten. In a subsequent contact with the same (or a chemically related) compound, the T-lymphocytes are now able to recognize the aggressor and, through the production of chemical mediators called lymphokines, cause the allergic reaction, such as erythema (redness) and edema (swelling), to occur. This process takes several hours to manifest itself, hence the name delayed hypersensitivity.

An essential step in this mechanism is the formation of a conjugate, a protein-hapten complex resulting from covalent bond formation. Because covalent attachment is involved in initiating ACD, some prediction of the allergenic properties of a compound can be made. To form a covalent bond, certain conditions must be fulfilled. Any protein can be considered as a nucleophile, because several functional groups of the constituent amino acids are nucleophilic (e.g., $-NH_2$, $-SH$, imidazole and, under favorable circumstances, even $-OH$). Haptens, on the other hand, can be considered as electrophiles, so that the hapten-protein conjugation can be envisioned as a nucleophile-electrophile interaction as shown schematically in Figure 4.

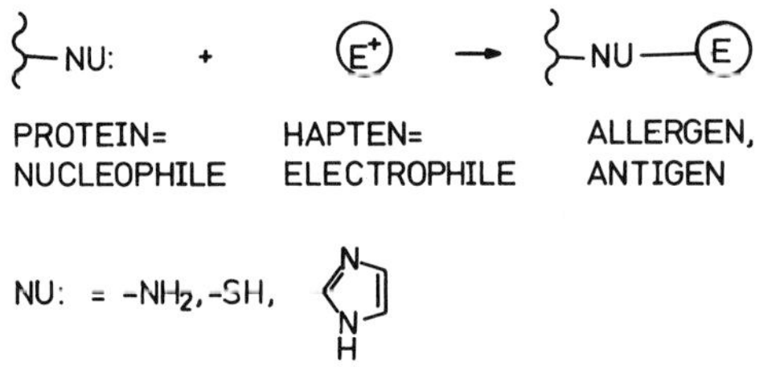

Fig. 4 Schematic representation of the hapten-protein conjugation reaction.

In principle, any molecule containing an electrophilic group can be a hapten, and the stability of the particular resulting protein conjugate will necessarily depend on the chemical nature of the hapten. Adducts with ketones or aldehydes are less stable than adducts with esters, lactones and Michael acceptors (Fig. 5). Michael acceptors are, in fact, excellent candidates for initiating ACD. It should be pointed out, however, that not all allergenic compounds are electrophiles. Some of them, prior to conjugation, must undergo a chemical transformation, such as oxidation, to become electrophilic. These compounds are called prohaptens. An example of a prohapten is the active principle of poison ivy, one of the most famous prohaptens in the United States. This plant contains an alkyl catechol; thus, binding to a protein implies preliminary oxidation to an *o*-quinone [10] (Fig. 6). Many halogenated bactericides are known to be photosensitizers (i.e., they become true haptens only in the presence of light). In this instance, it is thought that dehalogenation gives rise to a radical which then becomes attached to a cutaneous protein [5] (Fig. 7). Photosensitization has also been described for both psoralens (used in fragrances) and phenothiazinic compounds mentioned previously (Figs. 1 and 2). Coupling with a cutaneous protein in this instance also occurs via radical formation.

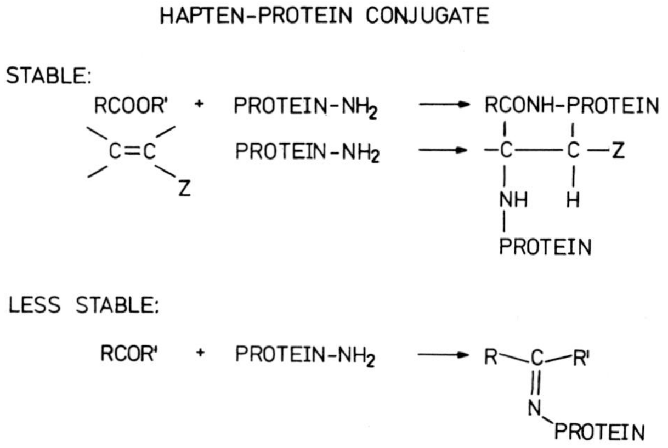

Fig. 5 A comparison of stable and less stable hapten-protein conjugates.

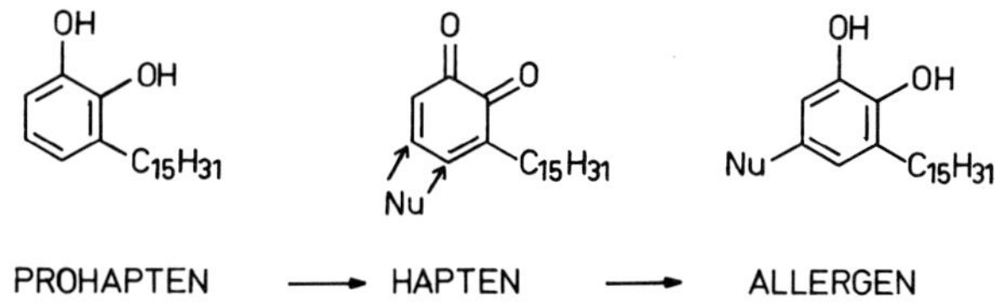

Fig. 6 The catechol prohapten of poison ivy must undergo oxidation to the *o*-quinone hapten before conjugation.

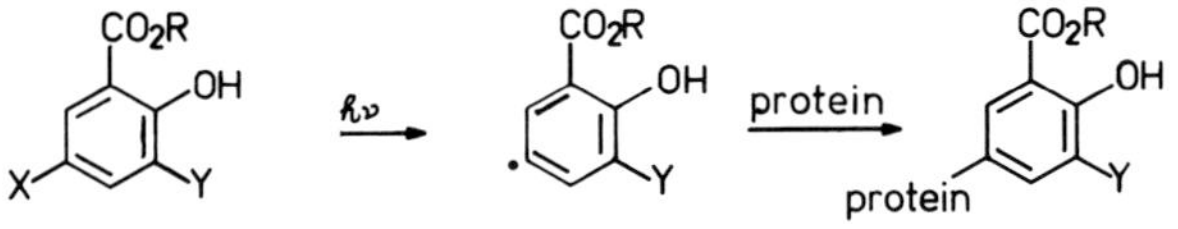

Fig. 7 Bactericides (x=y=halogen) undergo light induced dehalogenation to a free radical before conjugation.

Specificity of sensitization is rather broad, otherwise one would not have to worry too much about contact sensitivity, and simply avoiding the particular sensitizer would be enough to prevent dermatitus. However, when an individual is sensitized to chemical A, he can often cross-react to a second, chemically related, substance B. This effect is known as cross-sensitization or cross-allergy.

To experimentally study the sensitizing effects of a given substance, sensitization tests are conducted on man (the maximization technique of Kligman [11] is most frequently used) and on animals (usually guinea-pigs). Sensitization techniques for guinea-pigs are numerous and include the open epicutaneous technique, the patch test, the Freund complete adjuvant technique (FCAT, intradermal), the Landsteiner-Draize method, and the maximization technique [12]. The predictive value of these tests is extremely variable.

ACD to Oils and Essential Oils

In considering the potential and the known haptens among substances used in fragrances and flavors, it is first appropriate to describe known sensitizers among natural extracts and essential oils. These include costus root oil (haptens: costuslactone, dehydrocostuslactone), cassia oil (hapten: cinnamic aldehyde), cinnamon bark oil, Peru balsam (hapten: coniferyl benzoate), citronella oil, tansy oil, angelica root oil, laurel leaf oil, armoise oil and alantroot oil. In several of these oils the agent responsible for sensitization has been identified. Most of these haptens contain a conjugated system, ideally electrophilic for Michael-type addition (e.g., α-methylene-γ-butyrolactones and α,β-unsaturated aldehydes).

Functional Groups Responsible for ACD

Since haptens have been described as electrophiles (or potential electrophiles), it is possible to classify them in terms of their functional groups.

Alcohols. Peru balsam-sensitive patients cross-react to several unsaturated alcohols such as cinnamic alcohol (*1*) and geraniol (*2*) (Fig. 8). Very few saturated alcohols are responsible for allergic contact dermatitus, and among them are citronellol (*3*) (cross-reaction in citronella oil-sensitive patients), and myristic alcohol (*4*) in lanolin and ethanol (Fig. 8).

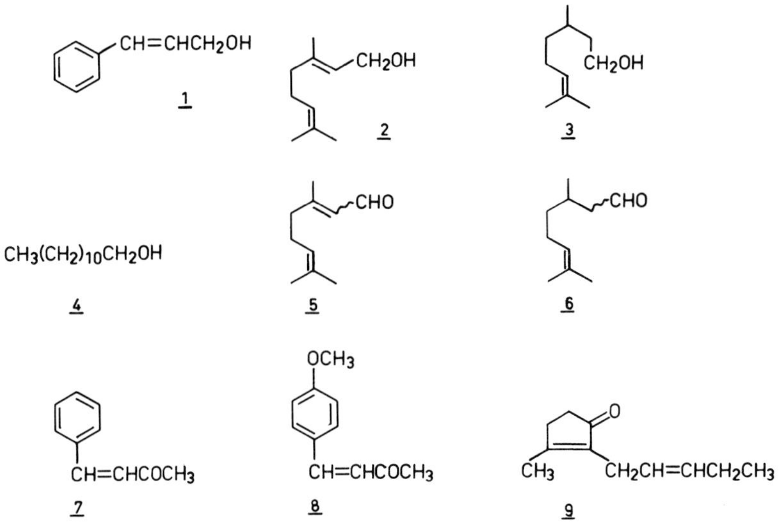

Fig. 8 Structures of cinnamic alcohol (*1*), geraniol (*2*), citronellol (*3*), myristic alcohol (*4*), citral (*5*), citronellal (*6*), benzylidene acetone (*7*), *p*-methoxybenzylidene acetone (*8*) and jasmone (*9*).

Aldehydes. Aldehydes, especially those containing α,β-unsaturation, are common sensitizers. The most important hapten with an aldehyde function is citral (*5*) (Fig. 8), a mixture of neral and geranial. A minor change in chemical structure can be responsible for complete loss of allergenic activity. Thus, citronellal (*6*) (Fig. 8) is almost never a sensitizer.

Ketones. There are very few known ACD cases to ketones; however, conjugated ketones such as benzylidene acetone (*7*) and its *p*-methoxy derivative (*8*) (Fig. 8) are allergenic. Jasmone (*9*) (Fig. 8), the synthesis of which has been described in this symposium (paper No. 8), could be a potential hapten because of the α,β-unsaturated system, although no ACD case has been described so far.

Esters. Esters are generally non-sensitizers, although they are sometimes irritants. Esters such as diethyl maleate (*10*) and acrylates (*11*) (Fig. 9) are, however, strongly allergenic.

Phenols. Phenols, particularly *o*- and *p*-diphenols (*12*) (Fig. 9), are allergenic. The most well-known allergen in the United States is urushiol, a mixture of four pentadecylcatechols (*13*) extracted from poison ivy. Resorcinols (*14*) (Fig. 9) are less sensitizing. This is easily understood, as only *o*- and *p*-diphenols are readily oxidized to electrophiles (quinones), thus becoming true haptens. Mono- and diphenols (Fig. 10) are also responsible for leucodermia (depigmentation or secondary vitiligo). Other known depigmenting agents are β-mercaptoamines (Fig. 10).

Hydrocarbons. Hydrocarbons are generally non-sensitizers. On the other hand, hydrocarbon autoxidation products, particularly hydroperoxides, are powerful haptens. This is the basis for the allergenic activity of Δ^3-carene (*16*) (Fig. 9); however, limonene (*15*), which is used in several fragrance products, is only rarely described as allergenic.

Lactones. Lactones occur frequently in natural products. A number of these compounds, such as sesquiterpene lactones, are known as potent sensitizers. The allergenic activity is due to the presence of an α-methylene-γ-butyrolactone group. Many plants, particularly the

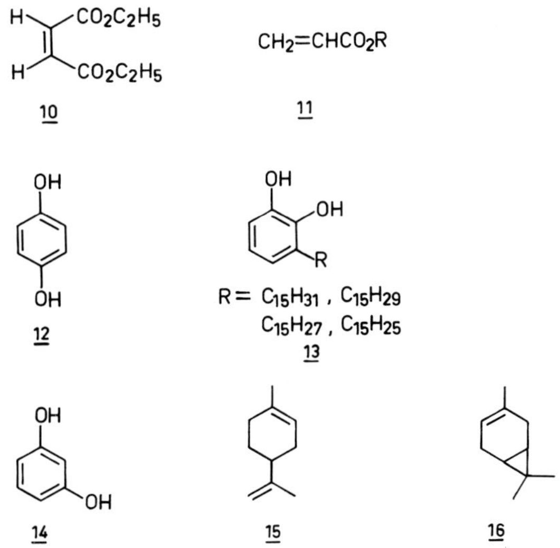

Fig. 9 Structures of diethyl maleate (*10*), acrylate esters (*11*), *p*-diphenol (*12*), urushiol (*13*), resorcinol (*14*), limonene (*15*) and Δ^3-carene (*16*).

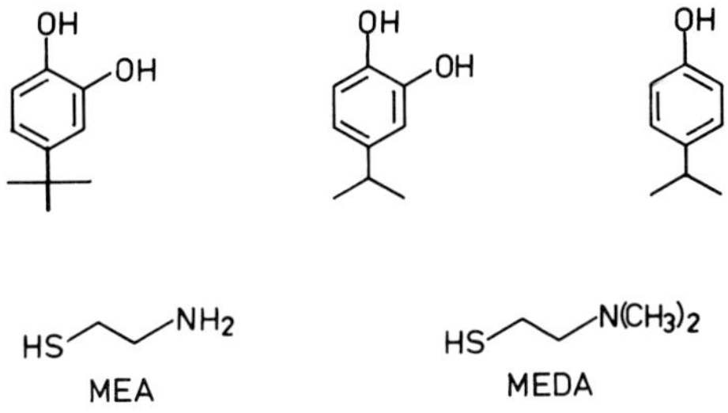

Fig. 10 Mono- and diphenols and β-mercaptoamines (mercaptoethyl amine and mercaptoethyl dimethyl amine) responsible for leucodermia.

Compositae, are responsible for a number of irritations, rashes and allergic contact dermatitus [13]. Extracts of many of these plants, for example camomile (*Matricaria chamomilla* L.), are being used in cosmetology. The sesquiterpene lactones responsible for ACD to these plants can be divided into four classes; germacranolides [e.g., costunolide (*17*)], eudesmanolides [e.g., alantolactone (*18*)], guaianolides [e.g., dehydrocostuslactone (*19*)], and pseudoguaianolides [e.g., damsin (*20*)] (Fig. 11).

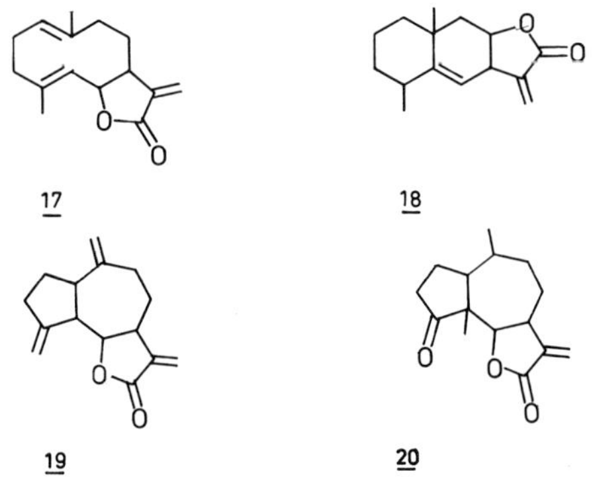

Fig. 11 Structures of costunolide (*17*) (a germacranolide), alantolactone (*18*) (a eudesmanolide), dehydrocostuslactone (*19*) (a guaianolide), and damsin (*20*) (a pseudoguaianolide).

In a chemical approach to understanding the mechanism of ACD, a number of α-methylene-γ-lactones were synthesized and tested [14–16]. The preliminary conclusion of this study is that, in order to give a cross-reaction in animals sensitized with alantolactone or methylene-γ-butyrolactone, a product must resemble the primary sensitizer. To determine if the binding of the hapten to a cutaneous protein was specific, ^{14}C-labelled lactones were synthesized and painted on the skin of a guinea-pig which had been sensitized to alantolactone (previous studies had shown the appropriate cross-reaction). The animal was then sacrificed, its skin and organs extracted, and the distribution of radioactivity determined. Most of the radioactivity remained in the "insoluble" skin, thus giving some support to the concept of peripheral sensitization of the T-lymphocytes.

Concluding Remarks

Prediction of the sensitizing properties of a particular compound is not very easy; however, there are some classes of compounds that should clearly be regarded with suspicion. Thus, the main hazardous problems associated with fragrance materials are Michael acceptors such as α-methylene-γ-lactones and unsaturated aldehydes, including cinnamic aldehyde and citral.

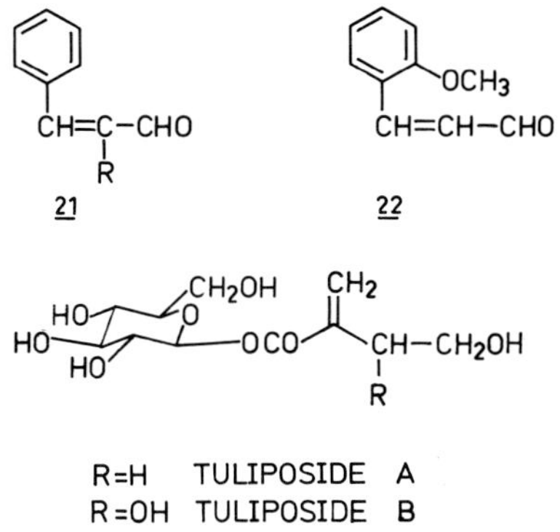

Fig. 12 Structures of α-alkyl cinnamic aldehyde (*21*), *o*-methoxycinnamic aldehyde (*22*) and the tuliposides.

One must, however, exercise some caution, as a change in the chemical structure can dramatically affect sensitizing properties. Thus, cinnamic aldehyde, which is often used in fragrances, is a potent sensitizer, but its α-alkyl derivatives (*21*) and *o*-methoxy derivative (*22*) are not (Fig. 12). A further illustration is provided by the hapten contained in onion bulbs; tuliposide A (R = H) is an active sensitizer, while the closely related tuliposide B (R = OH) is not (Fig. 12).

Acknowledgement

Financial support of the work described here by INSERM, DGRST and CNRS is gratefully acknowledged.

References

[1] Ducombs, G., and G. Chabeau, Dermato-allergologie de Contact, Masson, Paris, 1979.
[2] Foussereau, J., and C. Benezra, Les Eczemas Allergiques Professionnels, Masson, Paris, 1970.
[3] Steiberg, M., W. A. Akers, M. Weeks, A. H. McCreesh and H. I. Maibach, in H. I. Maibach, Animal Models in Dermatology, pp. 1–11. Churchill Livingstone, Edinburgh, 1975.
[4] McCreesh, A. M., and M. Steinberg, in F. N. Marzulli and H. I. Maibach, Dermatotoxicology and Pharmacology, pp. 193–210. John Wiley and Sons, New York, 1977.
[5] Haber, L. C., and A. R. Shalita, in F. N. Marzulli and H. I. Maibach, Dermatotoxicology and Pharmacology, pp. 427–440. John Wiley and Sons, New York, 1977.
[6] Maibach, H. I., and F. N. Marzulli, in F. N. Marzulli and H. I. Maibach, Dermatotoxicology and Pharmacology, pp. 211–224. John Wiley and Sons, New York, 1977.
[7] Harber, L., and R. Baer, J. Invest. Derm., **58**, 327 (1972).
[8] Coombs, R. R. A., and P. G. H. Gell, Clinical Aspects of Immunology, Blackwell Scientific Publication, Oxford, 1968.
[9] Polak, L., in F. N. Marzulli and H. I. Maibach, Dermatotoxicology and Pharmacology, pp. 225–288. John Wiley and Sons, New York, 1977.
[10] Mason, H. S., and A. Lada, J. Invest. Derm., **22**, 454 (1954).
[11] Kligman, A. M., J. Invest. Derm., **47**, 393 (1966).
[12] Klecak, G., in F. N. Marzulli and H. I. Maibach, Dermatotoxicology and Pharmacology, pp. 305–340. John Wiley and Sons, New York, 1977.
[13] Yoshioka, H., T. J. Mabry and B. N. Timmermann, Sesquiterpene Lactones, Tokyo Press, Tokyo, 1973.
[14] Benezra, C., G. Schlewer and J. L. Stampf, Rev. Franc. Allergol., **18**, 31 (1978).
[15] Schlewer, G., J. L. Stampf and C. Benezra, Can. J. Biochem., **56**, 153 (1978).
[16] Schlewer, G., J. L. Stampf and C. Benezra, unpublished results.

16. FDA'S RESEARCH AND REGULATORY PROGRAMS FOR COSMETICS AND FRAGRANCES

J. A. Wenninger*

Summary The research and regulatory programs for cosmetics and fragrances at the U.S. Food and Drug Administration (FDA) are described. Issues of current interest to the cosmetic industry, including nitrosamine and 1,4-dioxane contamination of cosmetic products and the fragrance ingredients acetyl tetramethyl tetralin (AETT) and 6-methylcoumarin, are reviewed from FDA's perspective. A general outline of FDA's regulatory authority and research programs for cosmetics is also provided.

Introduction

The safety of cosmetics and fragrances is coming under increased scrutiny by all segments of our society. Consumers are exposed daily to news media reports about potentially harmful substances in the food they eat, the drugs they take, and the cosmetics they apply to their body. It is unlikely that the debate over the safety of products will subside in the foreseeable future. What we hope will happen is that science will provide a basis for marking sound decisions in the public interest.

To understand the present environment under which cosmetics are regulated in the United States, one must be keenly aware of the limitations of FDA's legal authority to take action against harmful ingredients and products, and the limited resources available to do this job. In general FDA must show that a cosmetic product or fragrance preparation may be harmful under normal conditions of use, or that its labeling is false or misleading, before it can support a finding that a product is either adulterated or misbranded under the law. Adulterated and misbranded products are illegal, and may be subject to regulatory action. When the FDA does not have information about the safety of an ingredient or a product, it cannot take action. Since the law does not require manufacturers to supply information, the FDA frequently must develop the information or encourage others to do so. The major thrust of FDA's scientific programs for cosmetics centers around developing information about potentially harmful substances that may be components of cosmetic products.

* J. A. Wenninger, Food and Drug Administration, Division of Cosmetics Technology, 200 C St. S. W., Washington, D. C. 20204, U.S.A.

This review will focus on specific examples which are illustrative of the types of problems the FDA is faced with at the present time, the approach to dealing with these issues, and the policy the agency is following to improve the safety of cosmetic products.

Acetyl Tetramethyl Tetralin and 6-Methylcoumarin

Two substances of interest to the fragrance industry have recently been the focus of FDA attention. They are acetyl tetramethyl tetralin (AETT) and 6-methylcoumarin (6-MC). AETT has a potential for producing neurotoxic effects and/or generalized tissue discoloration in several species of animals [1]. 6-Methylcoumarin has been identified as a potent photocontact allergen (a photosensitizer) [2] (see also symposium paper No. 15).

The actions FDA took in regard to these two substances were quite different. Reportedly, both substances had a long history of apparent safe use in cosmetic products. In the case of 6-MC, the FDA alerted the industry to the problem and requested that cosmetic firms remove suntan or sunscreen products containing this ingredient from the market. In the case of AETT, the FDA relied on industry's voluntary discontinuance of the use of this ingredient. The reason for the different regulatory approaches was dictated by the type of information available, the limited authority under the law to obtain additional information, and what was believed to be the most appropriate use of limited resources.

In the case of 6-MC, the information available consisted of scientific studies demonstrating adverse effects of 6-MC in humans, expert opinion that 6-MC may be harmful under normal conditions of use, and documented case histories of consumer adverse reactions to products containing 6-MC [2, 3]. This information fully supported a finding that suntan preparations containing 6-MC were adulterated under the law. A major concern with continued use of 6-MC in suntan products was the possible induction of sensitization in a wide segment of the population not already allergic to the substance. In addition, the adverse reactions related to 6-MC in some products were not the usual transitory skin effects often associated with cosmetic allergic reactions. Some reports described blistering of the skin and systemic effects, that in certain cases required hospitalization of the affected person [4]. A recall of such products was the best course of action to prevent further consumer exposure to the harmful effects of 6-MC.

In the case of AETT, the only data available were animal test results [1], and the exposure levels under animal test conditions were significantly higher than those estimated for human exposure. Skin penetration data in humans were lacking, and it was not known precisely which products contained AETT. Industry's voluntary efforts to stop supplying and using AETT in cosmetics in the early part of 1977 affectively reduced consumer exposure to such products.

The neurotoxic effects of AETT certainly point out a new dimension which every fragrance supplier should now take under consideration when testing ingredients for safety. The FDA is carefully following studies on the neurotoxicity of AETT being carried out by Drs. Spencer (Albert Einstein College of Medicine) and Foster (Toxicology Department, Avon Products, Inc.) [1]. Significantly, it was the cosmetic industry that, during routine testing of AETT, discovered the blue discoloration of internal animal organs which led to the further testing that uncovered the neurotoxic effect of this compound. This information was shared with the FDA at a meeting requested by Dr. Opdyke of Research Institute of Fragrance Materials. This meeting served to make the information a matter of public record. It is our hope that similar situations in the future will result in similar cooperation between government and industry in the public interest. The FDA is planning to publish regulations which will prohibit the use of 6-MC and AETT in cosmetics and topical drugs.

Before leaving the subject of fragrances, one additional point should be mentioned. It concerns the traditional secrecy which surrounds the composition of fragrance preparations. The ingredients of specific fragrances are unknown to the public, FDA and most cosmetic firms that incorporate them into their products. Requiring the listing of fragrance ingredients on product labels has been suggested in the past, but not adopted by FDA because of the trade secret issues raised and the complexity of fragrance formulations. Eventually, a solution to the trade secret question will have to be worked out to better protect consumers from potentially harmful ingredients. A recent draft of a report to the President [5] discusses approaches to dealing with confidential information submitted to regulatory agencies about the composition of consumer products. It was interesting to note that this report recommends strengthened cosmetic legislation. One key element of the recommendation was the requirement for mandatory product formulation disclosure, which, presumably, also means fragrance formulations.

Cancer Causing Agents

Probably the most difficult issue facing FDA today deals with the presence of ingredients and contaminants in cosmetic products which have been demonstrated, by animal testing, to cause cancer. FDA's response to information that a substance directly added to cosmetics is a carcinogen has been to ban its further use in cosmetics. Chloroform and vinyl chloride are examples of ingredients banned from cosmetics because of their carcinogenicity. When authority under the law does not permit FDA to ban such ingredients, a warning statement for product labels has been proposed. Coal-tar hair dye ingredients fall into this category. Of special concern to FDA has been the recent discovery of two contaminants in cosmetic products which have been shown in animal tests to cause cancer. One of these substances is the nitrosamine, *N*-nitrosodiethanolamine [6, 7], and the other is 1,4-dioxane [8].

Nitrosamines

The contamination of cosmetics with nitrosamines or 1,4-dioxane may render a product adulterated under the law. However, the burden of proof that such a product would be harmful under normal conditions of use rests with FDA. With regard to *N*-nitrosodiethanolamine (NDELA), the FDA published a notice in the Federal Register on April 10, 1979 urging the cosmetic industry to take appropriate steps to reduce or eliminate, to the extent possible, NDELA contamination of cosmetic products. The FDA also advised that cosmetic products may be analyzed for nitrosamine contamination, and that individual products could be subject to regulatory action. As of June 30, 1979 the Division of Cosmetics Technology at FDA has analyzed 242 cosmetic products for NDELA. The results are given in Table 1. Additionally, the Cosmetic, Toiletry and Fragrance Association (CTFA) has sponsored contract studies on NDELA contamination of cosmetic raw materials. The latest CTFA results from the analysis of 70 cosmetic raw materials [9] indicated that 14 (20 %) were contaminated with levels of NDELA ranging from 50 to more than 1000 ppb (1 ppm).

Table 1. Summary of Results from the Analyses of Cosmetic Samples for *N*-Nitrosodiethanolamine[a]

Samples found to contain NDELA at:	Number	% of Total
Levels above 2000 ppb (2 ppm)	19	8%
Levels between 30 ppb and 2000 ppb (0.03–2 ppm)	58	24%
Trace levels 10 to 30 ppb (0.01–0.03 ppm)	17	7%
None detected	148	61%
Total number of samples analyzed	242	

[a] Analysis by Division of Cosmetics Technology, FDA, Washington, D. C. 20204, as of June 30, 1979. Most of the results have not been confirmed by gas chromatographic mass spectrographic analyses.

To date, the highest levels of NDELA were found in cosmetic products which also contain an amine and 2-bromo-2-nitropropane-1,3-diol (BNPD), which is used in some products as a preservative. BNPD has been recently shown [10] to be an active nitrosating agent. Under alkaline conditions, BNPD's decomposition products include both nitrite ion and formaldehyde. Formaldehyde probably catalyzes the nitrosation reaction under alkaline conditions. Therefore, information is now available which demonstrates that BNPD can be a nitrosating agent for secondary and tertiary amines under alkaline conditions. The Agency expects cosmetic firms to now take these factors under consideration when formulating cosmetic products.

In the most recent group of cosmetic products analyzed for NDELA, four products were encountered that contained both triethanolamine and 2-bromo-2-nitropropane-1,3-diol, but that contained no detectable level of NDELA. These are the first samples to contain the amine and BNPD in which NDELA was not found. The absence of NDELA in these products is of great interest, as detailed study of these products may reveal some important clues regarding inhibition of nitrosation reactions in cosmetic products.

Much work has gone into evaluating the adequacy of the methodology used to determine NDELA in cosmetic products. Most people working on this problem are in the process of modifying and revising the current methodology to improve its speed, reproducibility and accuracy. A substantial amount of work is being done by the cosmetic industry to find more economical and easily operated instruments for routinely determining NDELA. Uppermost considerations have been the avoidance of false positives, and the ability to reliably and independently confirm results obtained from the high pressure liquid chromatograph-thermal energy analyzer (HPLC-TEA). At this time, gas chromatography and mass spectrometry are being used for confirmation.

The FDA is in the final stages of validating revised methodology for isolating NDELA from cosmetic products before separation and determination by HPLC-TEA. So far, the recoveries of NDELA added to commercial cosmetic products have been satisfactory. The revised methodology does not involve the use of a silica gel column during isolation of NDELA from the product, as a substantial amount of NDELA can be lost with this technique. For this reason, it is believed that the results reported to date are lower than the true values.

The latest results from the analysis of cosmetic products for NDELA are now being reviewed to determine if regulatory action is warranted to protect the public from those products which contain this material. Each case will have to be evaluated on an individual basis, taking into consideration a number of factors which may affect human exposure to NDELA contained in a product. It has now been demonstrated [11] that application of an NDELA-contaminated cosmetic product to the skin can result in NDELA absorption by the body as detected by its appearance in the urine. FDA will also sponsor contract research studies to determine if there are *N*-nitroso compounds, other than NDELA, that may be contaminants of cosmetic products. On a theoretical basis, it is believed that there may be a wide variety of such compounds corresponding to the amines, amides, and other nitrogen containing raw materials used as ingredients in cosmetics. Since an estimated 80 % of the *N*-nitroso compounds tested are animal carcinogens, information concerning the presence of this class of compounds is needed to properly assess the health significance of this contamination problem.

1,4-Dioxane

Recently, 1,4-dioxane was found to be a contaminant of cosmetic raw materials [12], and this material has also been shown to be carcinogenic in animal tests [13]. The reported routes of exposure in animal tests included skin painting studies. 1,4-Dioxane should not be confused with the chemical dioxin, a highly toxic contaminant of herbicides that has attracted a great deal of interest.

1,4-Dioxane was first discovered by FDA in food-grade emulsifying agents known as polysorbates [8]. Polysorbates are also used in cosmetic products as emulsifiers. It is estimated that about one-third of the emulsion based cosmetics contain polyoxyethylene derivatives that may be contaminated with 1,4-Dioxane. 1,4-Dioxane can be formed as a by-product of the polymerization of ethylene oxide which is used to ethoxylate a wide variety of organic materials to form ethoxylated anionic, cationic and nonionic surfactants. These are widely used as emulsifiers, foaming agents and dispersants at concentrations ranging from less than 1 % to over 25 % in cosmetics.

The Division of Cosmetics Technology has recently completed its initial survey of ethoxylated cosmetic raw materials for 1,4-dioxane contamination, using analytical methodology developed at FDA which has a detectability limit of 0.5 ppm [8]. So far, 65 ethoxylated cosmetic raw materials have been analyzed for 1,4-dioxane (Table 2). From this limited survey, it appears that the contamination problem is rather widespread. The FDA is in the process of developing analytical methodology for the determination of 1,4-dioxane in finished cosmetic products, and also has studies underway to determine the skin penetration characteristics of 1,4-dioxane. These latter studies are being conducted with excised human skin and in monkeys.

At this stage, the level of risk posed by the use of cosmetic products contaminated with dioxane is uncertain. As in other cases involving carcinogens, the risk is related to the potency of the carcinogen, its skin penetration characteristics, and human exposure under conditions of customary use. The data on the carcinogenicity of 1,4-dioxane are now under review at FDA.

While the scope of this problem is under review, the Agency is urging cosmetic raw materials suppliers to analyze their ethoxylated ingredients for 1,4-dioxane. It has been reported [14] that some firms are modifying their manufacturing procedures for ethoxylated raw materials to

Table 2. Summary of Results from the Analyses of Ethoxylated Cosmetic Raw Materials for 1,4-Dioxane[a]

Samples found to contain 1,4-dioxane at:	Number	(%)
Levels above 100 ppm	8	(12%)
Levels between 10 and 100 ppm	23	(35%)
Levels between 0.5 and 10 ppm	22	(34%)
None detected	12	(19%)
Total number of samples analyzed	65	

[a] This summary based only on those samples analyzed by the Division of Cosmetics Technology, FDA.

reduce or eliminate 1,4-dioxane contamination. "Clean-up" procedures (vacuum stripping) are also being used to prepare raw materials that have no detectable levels of 1,4-dioxane (less than 0.5 ppm). From what is now known, it appears that the 1,4-dioxane contamination problem may be controllable by modification of both the manufacturing and clean-up processes for ethoxylated raw materials.

Concluding Remarks

In concluding this review, it is important to emphasize that scientific knowledge about potentially harmful substances used as ingredients, or that may contaminate cosmetic products, will increase dramatically in the coming years. While there are still some unanswered questions about the degree of risk such products pose to the public, the FDA is committed to minimizing this risk to the greatest extent possible through available administrative and regulatory authority.

References

[1] Spencer, P. S., A. B. Sterman, D. Horoupian, M. Bischoff and G. Foster, Neurotoxicology **1**, 221 (1979).
[2] Kaidbey, K. H., and A. M. Kligman, Contact Dermatitis **4**, 277 (1978).
[3] Holder, W. R., "Photoallergic Contact Dermatitis to 6-Methylcoumarin in Commercial Suntan Preparations; Cross Reactions to Coumarin", Unpublished manuscript of paper presented at the Texas Dermatological Society Meeting, October 29, 1978.
[4] Wenninger, J. A., unpublished reports (1979), Division of Cosmetics Technology, Food and Drug Administration, Washington, D. C. 20204.
[5] Report to the President by Toxic Substances Strategy Committee (August, 1979). Available from the Public Information Office (TSSC-Report), Council on Environmental Quality, Executive Office of the President, 722 Jackson Place, N. W., Washington, D. C. 20006.
[6] Fan, T. Y., V. Goff, L. Song, D. H. Fine, G. P. Arsenault and K. Biemann, Food Cosmet. Toxicol. **15**, 423 (1977).
[7] Federal Register **44**, 21365, April 10, 1979.
[8] Birkel, T. J., C. R. Warner and T. Fazio, J. Assoc. Off. Anal. Chem. **62**, 931 (1979).
[9] Winstead, J. A., CTFA Nitrosamine Task Force Report No. 1, May 30, 1979 (Cosmetic, Toiletry and Fragrance Association, 1133 15th St., N. W. Washington, D. C. 20005).

[10] Schmeltz, I., and A. Wenger, Food and Cosmet. Toxicol. **17,** 105 (1979).
[11] Edwards, G. S., M. Peng, D. H. Fine, B. Spiegelhalder and J. Kann, Toxicology Letters **4,** 217 (1979).
[12] Wenninger, J. A., unpublished results (1979), Division of Cosmetics Technology, Food and Drug Administration, Washington, D. C. 20204.
[13] Monographs on the Evaluation of Carcinogenic Risk of Chemicals to Man (1976), International Agency for Research on Cancer, Lyons France, **11,** 247–256.
[14] Greif, M., unpublished FDA notes from Cosmetic Ingredient Review Meeting, April 27–28 (1979).

17. FDA'S FLAVOR AND SPICE SAFETY REVIEW PROGRAM

M. V. Smith*

Summary FDA's regulatory authority to assure the safety of food flavoring materials is derived from the Federal Food, Drug, and Cosmetic Act and implemented through various regulations governing approval and use of ingredients which may be either Generally Recognized As Safe (GRAS) substances or food additives. The FDA's GRAS Review Program was initiated in 1972 and has been expanded to include substances regulated as food additives. Approximately 2000 flavoring ingredients are covered in this safety review. Scientific Literature Reviews including all published toxicological articles from 1920 to the present have been prepared on some 1280 synthetic flavors, and reviews are currently being compiled on the remaining natural flavors. Total dietary consumption of each flavor has been estimated by integrating industry usage, portion size, and frequency of eating data, or by simply using total poundage to derive per capita intake levels. Safety evaluation criteria, including minimal toxicological testing requirements, are discussed.

Introduction

Most papers of this symposium were concerned primarily with scientific issues. The regulation of flavors certainly involves the resolution of scientific questions and perhaps it would be a simple task if we had to deal only with these questions. However, as a regulatory agency the FDA must work within the framework of the law. When Congress drafts laws it is concerned with economic and social issues as well as the scientific basis for legislation involving technical matters. It is important, therefore, to first examine the statute upon which the agency must base all of its decisions and examine some of the problems of its interpretation.

The Federal Food, Drug and Cosmetic Act

The pertinent law is the Federal Food, Drug and Cosmetic Act of 1938, and more specifically the 1958 Food Additives Amendment to the Act [1]. This law requires that the safety and efficacy of food additives must be evaluated and approved by the FDA before they can be sold on the market. The amendment defines a food additive as any substance that becomes, or may

---

* M. V. Smith, Food and Drug Administration, Division of Food and Color Additives, Bureau of Foods, 200 C St. SW, Washington, D. C. 20204, U.S.A.

reasonably be expected to become, a component of food or otherwise affect the characteristics of food. A major class of substances which is exempted from this definition, and therefore is not required to have premarketing clearance, is comprised of those ingredients which are considered to be generally recognized as safe (GRAS) among experts qualified by scientific training and experience to evaluate their safety.

The exact meaning of GRAS status has been difficult to resolve. Congress provided minimal information in defining what criteria are to be used in determining GRAS status. The Act states that there are two bases for a GRAS determination: through "scientific procedures"; or through "experience based on common use in food" for substances used in food prior to January 1, 1958. FDA has expanded and clarified these criteria through the formal rulemaking process. General recognition of safety through "experience based on common use in food" must be based on generally available data and information. An ingredient not commonly used in food prior to January 1, 1958 can achieve GRAS status only through "scientific procedures". GRAS status through "scientific procedures" must generally be based on published studies. These studies must contain the same quantity and quality of scientific evidence as would be required to obtain approval as a new food additive.

The requirements for obtaining approval of a food additive are well-defined in FDA's regulations [2], and there is characteristically little dispute regarding their interpretation. The law clearly provides that the burden of showing safety is on the food additive petitioner, and he must satisfy the FDA before he can use his product. This burden of proof is not as well established when it comes to GRAS substances. The Food Additives Amendment does not specifically state how general recognition of safety is to be established, or who the qualified experts are to be. As a result, some industry groups have interpreted this law in ways which may vary significantly from FDA's interpretations. It is clear that the agency allows determinations of GRAS status which are completely independent of FDA. Although there are no legal requirements that GRAS affirmations must be obtained from the FDA before a flavor is marketed for food use, manufacturers are encouraged to submit such petitions for several reasons. The petitioner can take advantage of the scientific expertise and experience of the food safety experts within the Bureau of Foods. The industry can avoid substantial risk of loss of investments made in production and marketing programs and avoid any adverse publicity if FDA does not concur that the flavor is GRAS. Most important of all, the petitioner may avoid the public health hazard of marketing a product which is not GRAS, and, in fact, is harmful.

The FEMA Safety Review Program

The FDA has worked closely with the Flavor and Extract Manufacturers' Association (FEMA) to obtain information on the identity and safety of flavors and is continuing to do so in our current review. In 1960, FEMA determined that compliance with the Food Additives Amendment could be achieved by submitting all available information to a panel of independent and qualified experts who have complete freedom to decide which ingredients, in their judgement, are GRAS. The FEMA Expert Panel's conclusions have been published as a series of GRAS Lists beginning in October 1960 and most recently in July 1979 [3]. Thus far, FEMA has listed over 1600 natural and synthetic flavors as GRAS substances.

Early FDA Activities

Shortly after the enactment of the Food Additives Amendment in 1958, FDA requested FEMA to identify flavors which were commonly used in foods and were therefore eligible for GRAS status. FEMA identified 265 natural and 27 synthetic substances in this first list of flavors [4]. FEMA then sponsored a survey of most of the major manufacturers and users of flavors. The survey [4] revealed that more than 1400 flavors were being used in foods. These flavors together with those submitted in the form of petitions from Fritzche Brothers, Glidden-Durkee, the International Vermouth Institute, The Wine Institute and others form the basis for the FDA GRAS list of flavors and our list of flavors considered to be food additives [5]. The great majority of flavors that have been regulated as food additives or as GRAS substances by FDA are considered to be GRAS by FEMA. There is no difference between the requirements imposed on the conditions of use for flavors regulated as food additives and as GRAS substances, the only restriction being good manufacturing practice levels. The sole reason for regulating flavors as food additives was that FDA determined that there was no *general* recognition of the safety of these flavors because the data required to arrive at such a conclusion are not readily available to all safety experts [6].

FDA'S Gras Review Program

The so-called GRAS Review Program was begun by FDA in 1970 in response to a Presidential directive to the agency to review the safety of GRAS ingredients [7]. This directive was precipitated as a result of the cyclamate controversy. Under this program, an extensive literature search is made of all published articles containing safety information since 1920. This information is then examined and evaluated by the Select Committee on GRAS Substances of the Federation of American Societies for Experimental Biology (FASEB) under a contract with FDA [8]. To date FASEB has provided 100 final safety evaluations covering over 300 substances [9]. Eighteen more reports are expected in the next seven months to complete the review of all FDA published non-flavor GRAS substances. Although FASEB's recommendations are not binding, FDA carefully considers them and other information before it publishes its final actions in the Federal Register. These final rules either affirm the substance as GRAS for specific uses with any limitations, permit continued use of it under an interim regulation and require additional studies to answer the safety questions arising as a consequence of the review, or prohibit further use of the substance in food until additional studies are undertaken to demonstrate the safe use of the ingredient.

Only eight groups of flavors have been reviewed and evaluated by FASEB. They include caprylic acid; licorice, glycyrrhiza, and ammoniated glycyrrhizin; clove and its derivatives; allyl isothiocyanate, *p*-hydroxybenzyl isothiocyanate, and brown and yellow mustard; garlic and its derivatives; dill and its derivatives; oil of rue; and nutmeg, mace, and their essential oils [9]. All of these groups except nutmeg, mace and their essential oils have been proposed or finalized in the Federal Register as affirmed GRAS substances [10]. The tentative FASEB evaluation of nutmeg was received in February 1973.

The Select Committee concluded that "while there is no evidence that the levels of consumption (of these flavors) are capable of eliciting adverse effects, . . . the safrole and myristicin content of nutmeg products . . . should be further investigated to determine the range of content in both the East and West Indian products and to investigate or reinvestigate their possible mutagenic, teratogenic, and carcinogenic effects" [11]. If FDA strictly followed the FASEB recommendation, nutmeg, mace, and their essential oils would be regulated as interim food additives. Such action would require sponsors of these ingredients to undertake appropriate studies to resolve the questions raised about their safety, and periodic progress reports would be required.

It should be noted that the FDA received the FASEB safety evaluation on nutmeg over 6 years ago. The food safety issues which are directly confronted in this report are some of the most difficult facing the agency today. How should the FDA deal with natural flavoring substances which have a long history of use in food without any known deleterious effects but which are known to contain chemical components shown by themselves to be carcinogenic or in other ways deleterious? For example, safrole is known to be a natural component of cocoa, mace, nutmeg, black pepper, and anise, and studies have shown that continuous administration of safrole at 5000 ppm in the total diet of rats causes liver tumors [12]. The agency has banned safrole and many of its related substances from use in food. Additionally, foods which contain these substances (e.g., sassafras bark) and which are intended solely or primarily as a vehicle for imparting such substances to another food (e.g., sassafras tea) are banned [13]. Unresolved questions facing the agency include: what level of safrole in a natural substance would constitute a determination that the substance is used primarily for imparting safrole to another food? 20 ppm? 2000 ppm?; what kinds of tests can be used to determine this?; what is the scientific basis for allowing a substance containing 20 ppm to be used but not allowing one containing 2000 ppm?; and what is the legal basis for this in light of the Delaney clause?

Spices and natural flavors as a class have unique characteristics and present special safety evaluation problems in comparison to other substances (e.g., synthetic flavors) being evaluated by FDA. Natural flavors are chemical mixtures composed of innumerable constituents, many of which have not been identified. The presence and ratio of chemical constituents in these substances are a function of a variety of factors, including the age of the plant, growing conditions, the part of the plant from which the flavor is taken and the process used to prepare the flavor. These substances, therefore, may vary substantially from batch to batch. Also, the human exposure data currently available to the FDA as a result of the NAS Industry Use Surveys [14] may not be as accurate for spice and natural flavor use as the other substances surveyed, in that these flavors are typically added to food in the restaurant, bakery, or home rather than at the processing plant. Because of these unique problems, the FDA is currently considering ways of collecting and compiling the additional identity and exposure data required to assess the safety of natural flavors in a manner consistent with the review of other substances. However, until these policy issues are resolved and the agency can compile this required data on identity and exposure, the FDA will not be able to determine the appropriate criteria to be used in reviewing natural flavors and spices.

FDA's Cyclic Review Program

Excluding those substances already being evaluated in the GRAS Review Program, over 2300 other ingredients, including all the synthetic flavors, will be evaluted in what is known as the Cyclic Review Program [15]. Under this program, all toxicological tests and exposure data on the 1350 synthetic flavors known to be used in food in the United States will be examined. Scientific literature searches on the synthetic flavors have been compiled under contract with FEMA [16]. These searches cover all toxicological and safety data from 1920 to the present time. A plan has been developed for ranking these substances on a priority list based on exposure, structure, and known toxicity. The results of an initial pilot study indicate that nearly all of the 2300 compounds in this review will fail to meet the minimum standards of safety which have been deemed to be appropriate for a given level of exposure and a given molecular structure. The priority list will simply be a method of ordering these compounds in such a way that toxicological testing will begin first on those additives having the highest level of concern. Briefly, each additive will be scored on the amount of toxicological information known about it and will be placed in one of three levels, each requiring further testing if it fails to meet the agency's criteria. At level I, an LD_{50} test and a mutagenic screen will be required as a minimum. If positive results are indicated in the mutagenic screen, or adverse effects are observed in the acute study, additional tests will be required. Level II will require a mutagenic screen and subchronic feeding studies with two rodent species. Level III substances will require essentially the same testing needed for newly submitted food additive petitions, including two chronic feeding studies with *in utero* exposure. Not only will the safety criteria and priority list be published, but also the testing protocols. These protocols will allow the agency to make an objective evaluation of the quality of any previous toxicological tests and to describe what will be required in future tests.

Concluding Remarks

FDA's review of the safety of flavoring materials over the years has not been consistent, and is only recently becoming comprehensive. Agency concern over flavors has generally been low because of their low levels of use, the long history of use of many flavors, and their self-limiting characteristics. However, such low priority for these substances may not be totally justified in light of their reactive chemical nature and their potential for producing biological effects. As more is learned about the relative toxicity of flavoring materials in FDA's safety review programs, the agency will be able to adjust priorities for concern in order to assure the public health. The flavor industry should be prepared to undertake additional tests for some of these substances, and to work closely with FDA in providing any required safety data.

References

[1] The Federal Food, Drug, and Cosmetic Act, As Amended January 1979, 21 *U.S.C.* 301 ff. (1979).

[2] United States Department of Health, Education, and Welfare, Food and Drug Administration. Part 171-Food Additive Petitions, Title 21, *Code of Federal Regulations* § 171.1 (1979).

[3] Hall, R. L., *Food Tech.*, Oct. 1960; Hall, R. L., and B. L. Oser, *Food Techn.*, Dec. 1961; Hall, R. L., and B. L. Oser, *Food Tech.*, Feb. 1965; Hall, R. L., and B. L. Oser, *Food Tech.*, May 1970; Oser, B. L., and R. L. Hall, *Food Tech.*, May 1972; Oser, B. L., and R. A. Ford, *Food Tech.*, Jan. 1973; Oser, B. L., and R. A. Ford, *Food Tech.*, Nov. 1973; Oser, B. L., and R. A. Ford, *Food Tech.*, Sept. 1974; Oser, B. L., and R. A. Ford, *Food Tech.*, Aug. 1975; Oser, B. L., and R. A. Ford, *Food Tech.*, Jan. 1977;

Oser, B. L., and R. A. Ford, *Food Tech.*, Feb. 1978; Oser, B. L., and R. A. Ford, *Food Tech.*, July 1979.
[4] Hall, R. L., and B. L. Oser, *Residue Rev.* 24: 1-17 (1968).
[5] United States Department of Health, Education, and Welfare, Food and Drug Administration. Subpart F-Flavoring Agents and Related Substances, Title 21, *Code of Federal Regulations*, §§ 172.510-172.515 (1979) and Part 182 – Substances Generally Recognized As Safe, Title 21, *Code of Federal Regulations*, §§ 182.10–182.60 (1979).
[6] United States Department of Health, Education, and Welfare, Food and Drug Administration. Synthetic Flavoring Substances and Adjuvants. *Fed. Reg.* May 27, 1964, p. 6957.
[7] Nixon, Richard M., Special Message to the Congress on Consumer Protection, October 30, 1969, *Public Papers of the Presidents*, Mess. 421 (1969).
[8] United States Department of Health, Education, and Welfare, Food and Drug Administration, Select Committee on GRAS Substances. *Fed. Reg.* July 26, 1973, p. 20054.
[9] Select Committee on GRAS Substances, FASEB, *Federation Proceedings*, Vol. 36, No. II, p. 2527 (1977).
[10] United States Department of Health, Education, and Welfare, Food and Drug Administration, General recognition of safety and prior sanctions for food ingredients. *Fed. Reg.*, Dec. 7, 1976, p. 53614; *Fed. Reg.*, Dec. 7, 1976, p. 53616; *Fed. Reg.*, Aug. 2, 1977, p. 39117; *Fed. Reg.*, Aug. 26, 1977, p. 43092; *Fed. Reg.*, Jan. 27, 1978, p. 3704, *Fed. Reg.*, May 9, 1978, p. 19842; *Fed. Reg.*, Jan. 19, 1979, p. 3962.
[11] Select Committee on GRAS Substances, FASEB, Evaluation of the health aspects of nutmeg, mace, and their essentials oils as food ingredients (available from the office of the Hearing Clerk, Food and Drug Administration).
[12] Hagan, E. C., P. M. Jenner, W. I. Jones, O. G. Fitzhugh, E. L. Long, J. G. Brouwer and W. K. Webb, *Toxicol. Appl. Pharmacol.*, 7, 18 (1965).
[13] Unites States Department of Health, Education, and Welfare, Food and Drug Administration. Subpart B – Substances Generally Prohibited From Direct Addition or Use as Human Food, Title 21, *Code of Federal Regulations*, § 189.180 (1979).
[14] National Research Council, Committee on Food Protection. Subcommittee on Review of GRAS List (Phase II). A Comprehensive Survey of Industry on the Use of Food Chemicals Generally Recognized As Safe (GRAS). Washington: Natl. Acad. Sci., 1972.
[15] Anonymous, *Food Product Development*, Dec. 1979, p. 25.
[16] United States Department of Health, Education, and Welfare, Food and Drug Administration, GRAS or Prior-Sanctioned Direct Human Food Ingredients, *Fed. Reg.*, July 26, 1973, p. 20051.

18. CHANGING FOOD SAFETY PRINCIPLES APPLIED TO FLAVOR INGREDIENTS

R. D. Middlekauff*

Summary The manufacturers and users of flavor ingredients should anticipate that they will be expected to increasingly justify the need for flavor ingredients, the safety of these ingredients and their appropriate use in foods. This response by the flavor industry will cause a diversion of resources from the research now devoted to new products, new uses and new methods. The flavor industry must decide whether it will challenge the increasing regulations or cooperate fully with dedication of the necessary resources. Whichever course is followed with the research tools available, the industry is entering into an era marked by an extensive increase in understanding of flavor ingredients in all respects.

Introduction

As the numbers and types of all natural and synthetic flavor ingredients have increased over the years, so have the criteria for the evaluation of their safety as food ingredients changed significantly. The evaluation of the safety of flavor ingredients involves principles and procedures that have been, and continue to be, modified, improved, and otherwise adjusted on a nearly daily basis. The principles for evaluation are themselves complex, involving the interrelationship of many judgmental factors. In addition, the procedures, based as they are on rapidly improving scientific techniques, are also continually changing. Incorporated into a decision determining that a food or food ingredient is or is not safe, is the confrontation of many needs and interests. This shifting has created an unsettling situation for the flavor industry.

The factors which underlie the principles for safety evaluation need to be clarified, and accepted or modified as appropriate. Unless caution is used, the flavor industry may be forced into a course of activity which would be unnecessarily harmful to innovation and expansion, and which could even jeopardize continuation of the industry in its present form.

Importance of Flavors

Before resolving the question of whether or not flavor ingredients are safe for use in foods, it is important to determine if flavors are even important enough to be permitted for use in food regardless for their relative safety. If, by the standards of the Federal Food, Drug and Cosmetic

* R. D. Middlekauff, Bonner, Thompson, O'Connell, Gaynes and Middlekauff, 900 17th Street, N.W., Washington, D. C. 20006 U.S.A.

Act, flavors are shown to have some minimal degree of importance or benefit, the process of evaluation of safety should begin. This assessment of importance is applied with respect to each individual flavor ingredient upon a showing that it is a flavor ingredient.

Over the years, flavor ingredients have been placed in a special group, enjoying a somewhat privileged status. For example, in the United States, the Federal Food, Drug and Cosmetic Act exempts disclosing the name of flavor ingredients in food products. The implementing regulations of the Food and Drug Administration apply other special regulations to flavor ingredients. The FDA has established guidelines distinguishing between the natural flavors and the artificial flavors (grounded in economic considerations) on the basis that there is a difference in value between the two. The regulation specifically denies this label as signifying a distinction between the natural and artificial flavors on the basis of safety.

The distinction between "natural" and "artificial" has subsequently become somewhat obscured as the traditional methods for flavor creation are analyzed against the regulatory structure. Prior to the FDA's regulation, the methods of flavor development had a significance resulting in names grown out of tradition and common use. However, when the past methods of preparation are considered in light of new regulations, the results are confusing. What used to be placed into the natural category may end up being placed into the artificial category. As an additional consequence, new methods are developed, designed specifically to conform to the new regulatory language, resulting in what are now called natural flavors manufactured by ingeniously new methods. Throughout the application of the sorting process, the question of safety has not entered into the picture, although it could be a consideration in the future.

Applying creative ingenuity to a reasonably fixed number of flavor ingredients, the flavor industry has blossomed over the years with a continued growth in numbers and varieties of flavor mixtures, and with a continued growth in use application. The needs for more and different flavor mixtures have produced a growing, developing industry. The past and the future growth of the flavor industry is evidence of the importance of flavors. The demand for flavors is not simply an illusory demand, it is a real demand to serve a real need. The flavors are ingested not for themselves, but for what they contribute to the product in which they are incorporated. If the food product serves a need, then the flavors' importance is also confirmed.

Undoubtedly, the full significance of flavors in food has yet to be determined. Flavors, on ingestion or even before, may cause physiological activity which encourages ingestion and aids in digestion. Flavor ingredients also have the effect of permitting otherwise unpalatable food to be acceptable. While in some circles that may be considered as an undesirable trait (a concealment of unwholesomeness or of spoiled food, etc.), in other circles that is an advantage. Recent studies have indicated that the available land for animal and food use is shrinking every year. By the year 2000, the world's present population of 4.4 billion is expected to reach 6.5 billion. If, in addition, the developing areas of the world achieve minimum dietary levels, then the demand for food may be doubled by the year 2000. The present population of the United States to expected to increase from 212 to 260 million by the year 2000. All this will cause an increased demand for U.S. farm production of nearly 50%, but the productivity per acre probably will not be increased by more than 30%. Despite this growing need, land is being lost each year because of air pollution, soil erosion, water pollution and other environmental problems, as well to urbanization and other noncrop uses. This loss has been estimated at 1 million acres per year [1]. If the use of flavors will alleviate the impending crisis in the nation's and the world's food supply, by making edible otherwise unpalatable foods, then foresight of the regulatory agencies in considering such benefits is essential [2]. In the context of the world's food supply, the use of artificial or synthesized flavors, in lieu of natural flavors, has an additional value since they do not require arable land for their preparation.

Regulatory Impact

During the past two hundred years in the United States, a clearly discernible trend of regulation has developed. More agencies have been created during the 1970's than during any other ten year period of our government. The number created during the 1970's may even exceed the total number of federal agencies created during all the years prior to that period. The consequences of that trend are as expected, a shift from selfregulation to government-imposed regulation. Industry is finding itself in a partnership with the government. Not only does industry split its profits with the government, it also must allow the government a significant voice in the activities of the industry. In this trend, the activities of the Food and Drug Administration are no exception.

During the FDA's development as a governmental body, each year has been marked by more complex and more restrictive regulations. During the 1970's the enforcement activities have remained fairly constant, but Title 21 of the Code of Federal Regulations has been increasing by approximately 80 pages a year, or 57,000 additional words to be considered. The FDA has continued to expand, not only with the support of the public but also, and more importantly, with the insistence of the public. Loud voices have told the FDA, directly and through the offices of Congress, to do more to protect the food supply and other consumer products within their jurisdiction. Consumer advocates have had their influence on food legislation and they have demonstrated an effective influence on Congress. In addition, they have found legal actions beneficial to their goals. They have even been placed in important government agencies, with the surprising result that they are finding that being a government employee is not as easy as previously thought.

A recent study for the Department of Commerce [3] came to the conclusion that, "The regulatory agencies have immersed the country in a relatively impenetrable, incomprehensible and unmanageable morass of regulations." The report stated emphatically that the federal regulatory programs need improvement, since the consequences of over-regulation are several-fold. Over-regulation causes a diversion of capital investment to non-productive areas, an increase in the cost of product development, a lengthening of the product development cycle, an addition of uncertainty to decision-making, an inadequate protection of trade secret information, a lowering of industrial research and development expenditures, a diversion of critical scientific and engineering talents into defensive projects and away from creative products, and a construction of absolute barriers to entry [3]. In response to these assertions, public interest groups take the position that industry is willing to spend more for advertisement than research and development, so research and development really must not be all that important. Furthermore, they say that research and development for reasons of safety and health should not be considered non-productive. Industry must simply adapt to the new political and scientific realities [3]. To summarize their view, corporations today have an obligation to consider the human, environmental and social effects, and must deal with these issues that prompt governmental action, then the government won't be forced to act.

One of the key examples of the food industry's concern regarding over-regulation, is that the FDA requires 18 to 24 months, on the average, to approve food additive petitions. While the regulators apparent caution in making decisions on scientifically grounded determinations is, in part, due to the understandable difficulty in adjusting to advances in analytical techniques and to the detection of absolute zero concepts, the consequence of such regulatory approval delays is an erosion of financial incentives to innovators. The FDA is further hampered since it develops regulations without external communications, thereby excluding from its deliberations those who would be able to provide them with vital necessary information. Of greatest significance, however, is the fact that, as the safety of food ingredients is challenged, there is a growing diversion of researchers and resources from innovative research to technical compliance and defensive research.

Food and Drug Administration

Over the years, the number of bills introduced in Congress which affect the FDA have steadily increased. Of the 17,125 bills introduced during the 95th Congress (1977/78), 919 affected the FDA [4]. Fifty-two percent of those were product-related, and the remainder were reform-related (i.e., improvement of administrative procedures). In 1978, Congress held ten hearings on food safety and during 1973–1978, the most common subject of Congressional letters to the FDA was the matter of food additives. Congress will continue to study food safety in even greater depth, having released a comprehensive study on food safety [5] and proposed legislation [6].

The FDA's budget reflects the relative importance placed on its many activities. For 1978, the total budget was 295 million dollars, and 34 % was spent in food-related activities. The food budget was divided as follows: 19 % devoted to food additives, 18 % for food contaminants, 48 % for food sanitation and food processing surveillance, and 15 % for all other food-related activities [4]. Another view of the relative importance of the areas within FDA's jurisdiction is obtained from the results of the FDA's 1978 Consumer Food Label Survey [7]. Participants were asked to respond to the following: "Aside from prices, please tell me about any particular problems, difficulties, or concerns that you have with food these days." A surprising group of 49 % stated that they had no concern other than price. Another 14 % were concerned with product quality, and 8 % were concerned with freshness. The remaining 14 % were concerned about possible problems with adverse health implications of food ingredients. Those 14 % were asked to be more explicit about the nature of their concerns, with the result that 35 % were worried about preservatives, 29 % about additives or chemicals in general, 20 % about sugar content, 8 % about artificial colors, 7 % about salt content, and 4 % about artificial flavors. Thus, only 4 % of the 14 % interviewed were worried about artificial flavors. Apparently none of those interviewed were worried enough about natural flavors to specifically mention that concern.

In the broadest sense, FDA's Bureau of Foods' basic concern is the impact of food and food ingredients on man as toxic elements. Development of guidelines for distinguishing between safe and unsafe ingredients has been difficult and prolonged. With respect to safety considerations, it would appear that flavor ingredients are regulated no differently than any other food ingredients. The same concept of "generally recognized as safe" and the same food additive regulatory guidelines apply. The factors which traditionally enter into the question of safety determination include the specifications of the substance, its level of use, the consumption pattern of the foods in which it is used, effects found in feeding studies, and application of the appropriate safety factor. Of course, there are reasons why flavor ingredients should be treated differently. One of these factors has already been identified, namely, the importance of flavors. The fact that flavor ingredients are used at minute levels in food is another consideration (only a few milligrams or fractions of a milligram of any individual flavor compound is consumed by a person in one day). Low level of use alone is not, however, persuasive (i.e., pesticides are used at extremely low levels also).

The special position of flavor ingredients stands to be eroded as it meets the increasing demands of the regulators. When the FDA regulates a food ingredient, the principal issue is whether to allow the use of the ingredient and at what level it should be permitted. In the case of flavor ingredients, a third factor may be introduced, namely, whether the name of the ingredient should be indicated on the label. The relationship of safety and toxicity to the disclosure of ingredients is a growing matter of interest. In many cases, the nature of the ingredient is important to the consumer who is endeavoring to avoid consumption of that ingredient (for reason of allergy, sensitivity, etc.). How does this affect flavor ingredients? Of first interest is the level of use of flavor ingredients. It would require a fairly toxic ingredient to be of concern, since seventy percent of all flavor ingredients are used at less than 1000 lb/year in the U.S. [8]. Only a very few flavor ingredients have been identified which have raised a question of toxicity sufficient for removal.

The label of a bulk flavor product (material shipped from a flavor manufacturer to a food processor) must identify any flavor ingredient which is not permitted by a regulation of the FDA, or is not on an approved industry list. Even though that requirement does not constitute a limitation in the use of a flavor ingredient, you will probably not find any flavor ingredients listed on bulk flavor packages. Is this because the flavor manufacturer does not want to disclose a trade secret that he may have, namely, the use of that particular ingredient, or is this because he is concerned that the food producer will not purchase a bulk package when the package contains a flavor ingredient which is not permitted by the FDA in a regulation or is not on an industry list? Suppose the FDA decides that certain flavor ingredients must be listed on the ingredient statement of a food product, and appropriate authority is provided by legislation. What effect would this have on the use of those ingredients? The "Nutrition Labeling and Information Amendments of 1979 to the Federal Food, Drug and Cosmetic Act", recently introduced by Senators Kennedy, McGovern and others, would authorize the FDA to require "that the common or usual name of a flavoring be on the label for the purpose of providing health information to consumers" [9]. Would this be, in effect, a ban of those ingredients? Oddly enough, it may, even though the requirement is nothing more than placing that flavor ingredient in the same category as every other non-flavor ingredient. To take this one step further, suppose the consumer activists will be able to convince the FDA and Congress that all flavors and spices should be specifically identified on the label of a finished food product. Would this be preferable than to be in the position of having to declare only one or two of the flavor ingredients? The flavor industry will be subjected to these pressures and it must be prepared to justify its position that it is necessary to have many different flavor ingredients, that it is necessary to have artificial flavor ingredients, and that it is necessary not to disclose the identity of all flavor ingredients. On top of these burdens, the industry must be prepared to respond to the continuing and increasing demand for safety information on food ingredients, including flavor ingredients.

The criteria for evaluation of the safety of flavor ingredients has undergone the same scrutiny as the criteria for evaluation of the safety of food ingredients, but with a result balanced and responsive to the particulars of the flavor industry. The Expert Panel of the Flavor and Extract Manufacturers Association established its criteria as early as 1959 (Table 1). Following an assignment by the FDA for a recommendation, the Federation of American Societies of Experimental Biology produced its own suggestions in 1976 based on three levels of evaluation procedure [8] (Figs. 1, 2 and 3). The Food Safety Council proposed in 1978 a Decision Tree

Table 1. Summary of Criteria used by F.E.M.A. Expert Panel

1. Analogies with chemical-related substances with known toxicity or metabolism
2. The nature, level and volume of use of the substances in food
3. The toxicological significance of the levels of use
4. Available toxicity data
5. Metabolic pattern of compound in the body
6. Occurrence of the substance in natural foods

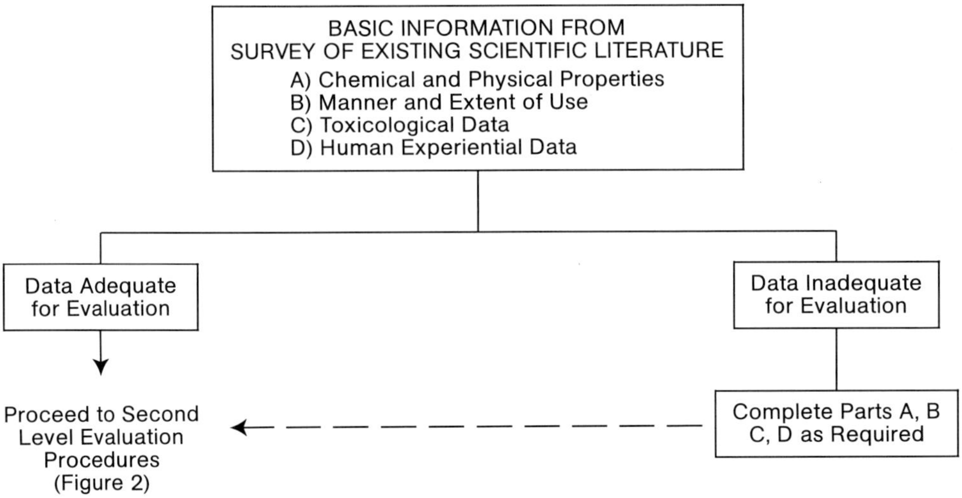

Fig. 1 First Level Evaluation Procedures.

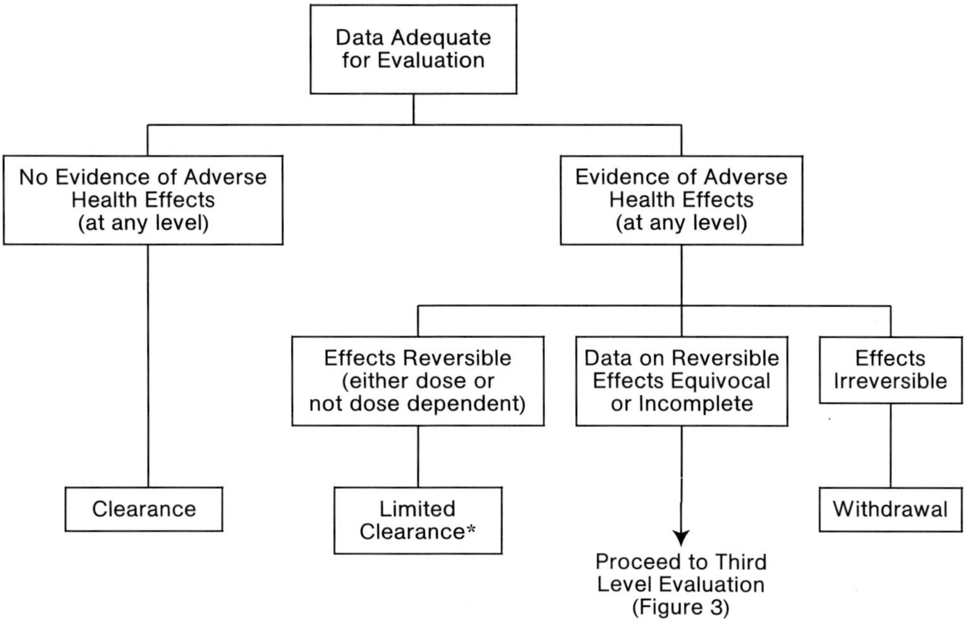

Fig. 2 Second Level Evaluation Procedures.

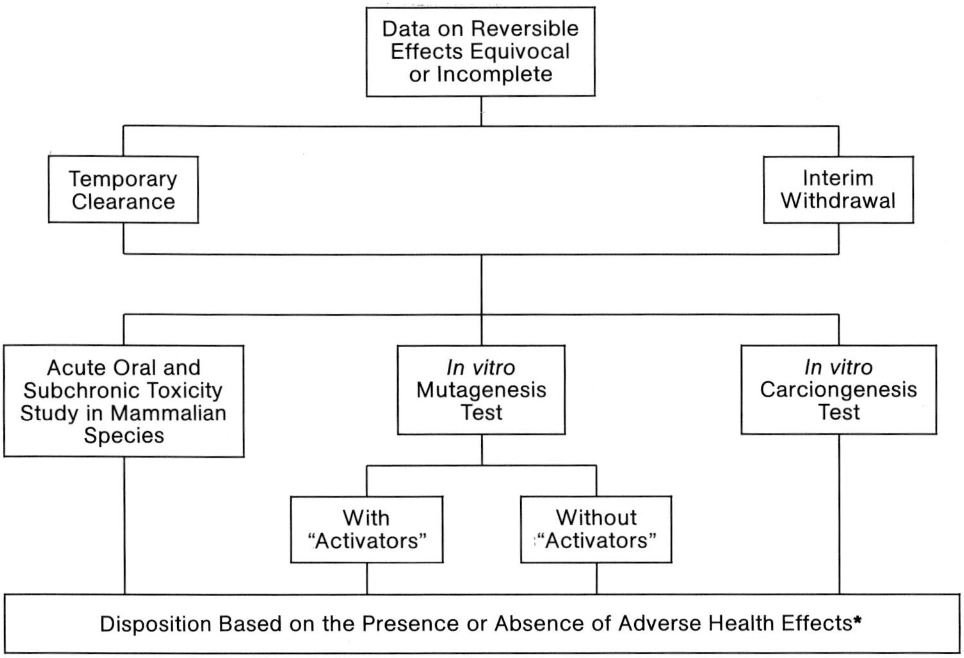

Fig. 3 Third Level Evaluation Procedures.

approach applicable to all food additives [10] (Fig. 4). The National Academy of Sciences put its oar into the water in 1979, and its proposal includes innovative suggestions in the evaluation of food additives, such as cost, a balance of benefits and risks, and the use of warnings as a viable alternative [2]. In an effort to establish a priority format, to assist in determining the sequence, scope and intensity of investigative effort so as to avoid an equivalent level of testing for virtually every substance, FEMA prepared a Decision Tree [11]. The Decision Tree classifies

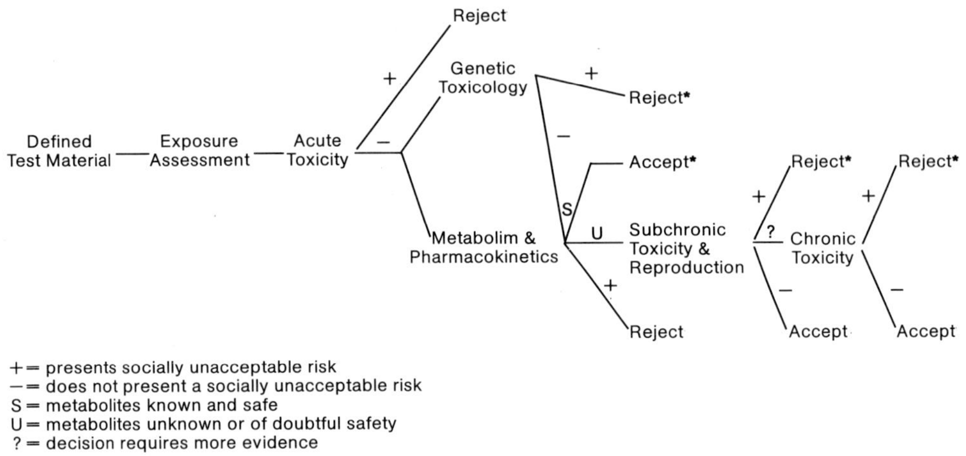

Fig. 4 Safety Decision Tree Proposed by the Food Safety Council [10].

substances according to a presumptive toxicity which, when combined with human intake, results in a basis with which to establish priorities of testing, and conceivably, the extent of testing. The FDA will propose and eventually adopt its own criteria, based hopefully on the best of the developments by the other groups, and with careful regard to the special aspects of flavor ingredients.

The Federal Food, Drug and Cosmetic Act has been described as "complicated, inflexible, . . . inconsistent in implementation . . . and inadequate to meet changing and increasing problems of food safety . . ." [2]. The FDA is finding itself in the position where it is asking Congress to be more explicit in its requirements [5]. A recent speech by an administrative official identified policy considerations for changes in the food safety laws [12]. The suggestions he considered included the categorization of foods and food ingredients as follows.

1. Traditional foods: This category would include basic food commodities, natural spices and flavors (and artificial spices and flavors that are chemically indistinguishable from natural ones), and essential nutrients at the levels they occur naturally in food. The Delaney Clause would not apply. The government would have the burden of showing significant risk to consumers before a substance could be banned.
2. Food contaminants: This category would include environmental and accidental contaminants. The Delaney Clause would not apply. The burden of proof would be on the government to prove potential risk to consumers, before setting tolerances or limitations.

3. Food additives: This category would catch every other food ingredient now regulated as food additives, GRAS, and color additives, plus any others not within the preceding two categories. There would be pre-market approval and a burden on the manufacturer to demonstrate safety. A modified Delaney Clause would apply. The proposal would permit scientific descretion rather than impose an absolute prohibition.

Finally, his proposal would permit the balancing of health risks. While there may not be universal agreement regarding these suggestions, they represent a substantial improvement over the existing confusion, and they provide a basis for further constructive discussion.

Looking Ahead to the Future

What should be done to encourage innovation in the food industry, rather than to have over-regulation stifle growth? The current direction of the Bureau of Foods suggests that industry develop and use the safest food substances or use more sophisticated processing control, with technology in place of chemical substances. The Director of that Bureau asks [5] that industry assist in reducing scientific uncertainty now associated with decisions, and collect information regarding normal background hazard from present food supply. He added that better methods are needed for quantifying risk and measuring biological phenomena such as behavior.

The FDA could take further steps. It could continue to indicate those areas where improvement is desirable and provide incentives to achieve in those areas [13]. It could speed up the review process by encouraging expeditious treatment and responses. Perhaps a limited use of a substance could be permitted while testing is in process, with additional allowable uses added on as testing progesses. It could establish priorities for regulations and administrative actions according to real needs, rather than in response to what may be termed frantic demands of consumer activists.

The National Academy of Sciences, in its study of a Food Safety Policy at the request of Congress, made the following three recommendations [2].

1. "that there be a single policy applicable to all foodstuffs, food additives, and food contaminants, and that the public official responsible for implementation of that policy be given sufficient flexibility to factor risks, benefits and other considerations into account when making a decision concerning a material that has been called into question"
2. "that there be available to that official options other than decisions simply to ban or not ban"
3. "that to facilitate such implementation, materials under consideration first be categorized as exhibiting low, moderate or high risk"

During the past few years, as the sciences of toxicology, pathology, nutrition, and related expertises have developed, regulatory officials have broadened their effort to respond to the health needs of the American public. No longer are these concerns identified in simple traditional terms; they have now been expanded to include concern about the impact of food additives on behavior and worry about "insidious chronic effects which rob us of our health and diminish the quality of our lives" [5]. Yet, the FDA needs to establish priorities–to emphasize consumption of food additives as being the source of hazards in the food supply is misleading to the American public. First, the public needs to be educated regarding their own nutritional requirements, based on the results of effective research as to nutritional needs. Once the public's diet is thereby made more predictable and more rational, the aberrant consumption patterns would no longer be a focal point, conceivably eliminating much of the concern for safety of food additives. Further research is necessary into the types of foods and consumption patterns that result in the degenerative illnesses prevalent today, such as, heart disease, some cancers, stroke, hypertension and diabetes [14]. The Congressional Office of Technology Assessment has found that the Federal Government has failed to adjust the emphasis of its human nutrition research activities to deal with the changing health problems of the people of the United States [14]. The American public seems to have misplaced its concern; it is accusing food additives for ills that might more appropriately be attributed to overconsumption, lack of exercise, smoking, improper diet, and other factors. The public does not understand that foods, not food additives, are the villains, because of the consumption and the extent of consumption of the food [15]. If these steps are followed, the safety of flavor ingredients will be judged in a rational, tempered climate, suitable for continued growth of the industry and supportive of innovation.

References

[1] Agricultural and Environmental Relationships: Issues and Priorities, Report prepared for the Subcommittee on Natural Resources and Environment of the Committee on Science and Technology and the Subcommittee on Conservation and Credit of the Committee on Agriculture, U.S. House of Representatives, 96th Congress, 1st Sess. (1979).

[2] Food Safety Policy: Scientific and Societal Considerations, A Report Prepared in Response to Public Law 75–203 (Saccharin Study and Labeling Act, National Academy of Sciences) pp. S–5, 4–12, 5–7 (1979).

[3] Report of the Industrial Advisory Committee to Department of Commerce (1979).

[4] Annual Report of the Food and Drug Administration (1978).

[5] Miller, S. A., Achieving Food Safety Through Regulation, Speech at Institute of Food Technologists Annual Meeting (June 1979).

[6] S. 1651 and S. 1652, 96th Congress, 1st Session.

[7] Consumer Food Label Survey, Food and Drug Administration (1978).

[8] Criteria for Evaluation of the Health Aspects of Using Flavoring Substances as Food Ingredients, Federation of American Societies for Experimental Biology, p. 12 (1976); See also comments filed by Expert Panel of the Flavor and Extract Manufacturers' Association on April 14, 1977, FDA Docket regarding 42 F.R. 10065–66.

[9] Sec. 3, S. 1652, 96th Congress, 1st Session.

[10] Proposed System for Food Safety Assessment, Report of the Scientific Committee of the Food Safety Council, p. 9 (1978).

[11] Cramer, G. M., R. A. Ford and R. L. Hall, Fd. Cosmet. Toxicol. **16**, 255 (1978).

[12] Remarks by Thomas P. Grumbly, Associate Administrator, Food Safety and Quality Service, U.S. Department of Agriculture, before the Food Animal Industry Conference, American Farm Bureau Federation, Chicago, IL (September 18, 1979).

[13] Department H.E.W. Memorandum, from Commissioner and Deputy Commissioner of Food and Drug Administration to Policy Board (June 29, 1979).

[14] Nutrition Research Alternatives, Office of Technology Assessment, Congress of the United States (1978).

[15] Consumer organizations rank evaluating the safety of food as a high priority item, but the FDA's Bureau of Foods ranks it sixth out of its eight programs. Discussion of Priorities: Fiscal Year 1980 Planning Process, Food and Drug Administration (April 1978).